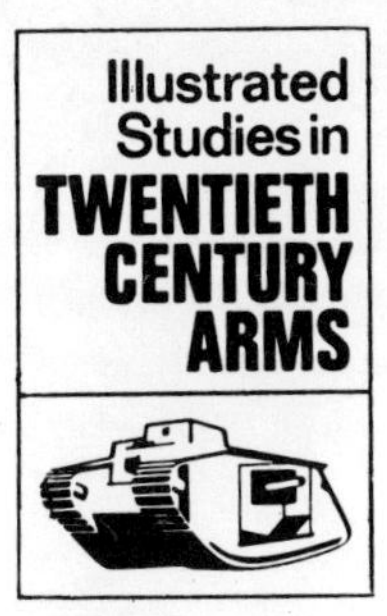

GERMAN INFANTRY WEAPONS
OF WORLD WAR II

Other volumes in this series

BRITISH AND AMERICAN INFANTRY WEAPONS OF
WORLD WAR II by A. J. Barker

TANKS OF WORLD WAR I: BRITISH AND GERMAN
by Peter Chamberlain and Chris Ellis

ALLIED BAYONETS OF WORLD WAR II
by J. Anthony Carter

1. *German Infantry in Action.*

GERMAN INFANTRY WEAPONS OF WORLD WAR II

by

A. J. BARKER

ARMS AND ARMOUR PRESS

Published by
Arms and Armour Press
Lionel Leventhal Limited
677 Finchley Road
Childs Hill
London, N.W.2

© A. J. Barker 1969

S.B.N. 85368 013 2

Plates 11, 20 and 38 are reproduced by courtesy
of the Imperial War Museum, London

Printed in Great Britain
by Photolithography
Unwin Brothers Limited
Woking and London

Contents

2. 9mm "08" Luger pistol.

GERMAN INFANTRY WEAPONS
OF WORLD WAR II

With the Treaty of Versailles, limits were placed on the size of Germany's armed forces, and the German Army was restricted to 100,000 professional soldiers. Heavy tanks and other sophisticated weapons of war were specifically excluded from its equipment, and the German General Staff was supposed to be abolished. Had this latter clause really been carried out, the German army would have been decapitated and deprived of its brain. But the leaders of the new *Reichswehr* decided to retain secretly the essentials of the old General Staff organisation. By various subterfuges, the foundations necessary for a rapid expansion of the army were laid; and when the abolition of the last prohibiting paragraphs of the Treaty of Versailles were proclaimed in 1935 the German army was well prepared. Close studies of foreign tactics, strategy and armaments had continued as in the time of the Kaiser, and with the connivance of German industry prototypes of guns, machine guns and tanks had been developed. Development continued throughout the war with consequent changes in composition of armament of Germany's armed forces, and the object of this book is to record the progress of development in German infantry weapons until their cessation in 1945.

By 1943 it was already clear that the trend in development was towards smaller, mixed units with greater fire-power, and the allotment of more and more automatic weapons to every unit.

The Germans appear to have been fascinated by small arms development, and there can be little doubt that the work done by their engineers and designers has had far-reaching effects on the equipment of the post-war forces of other nations.

THE GERMAN INFANTRY

Infantry formed the bulk of the German army. In 1940, after the *Blitzkrieg* in France, it seemed as if infantry had been replaced in importance by the *Panzer* army. But the war in Russia soon showed that this was not so, and the need for more infantry with greater fire-power to counter the numerical superiority and tenacity of the Russians was soon apparent.

The basic infantry unit of the German army was the *Regiment*—roughly equivalent to a British brigade or an American regiment.

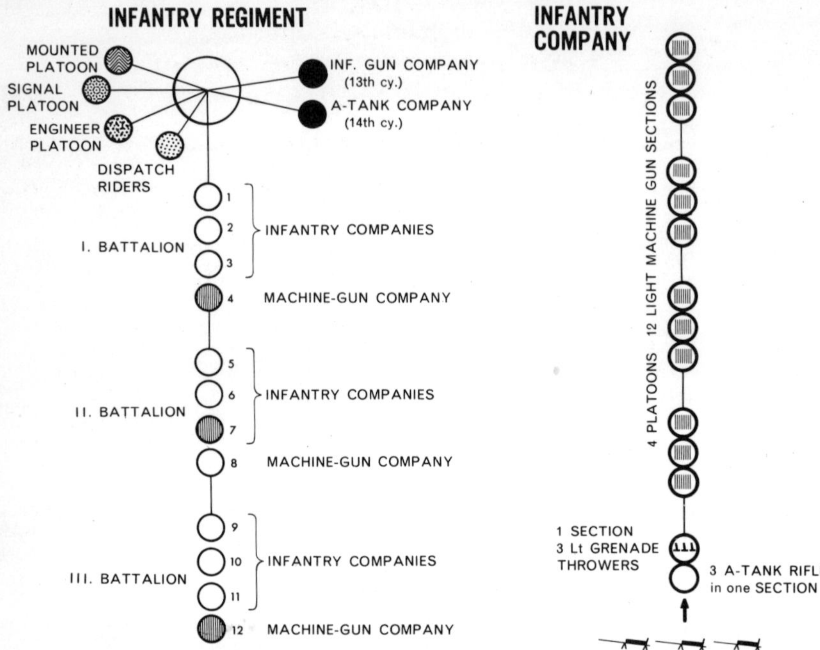

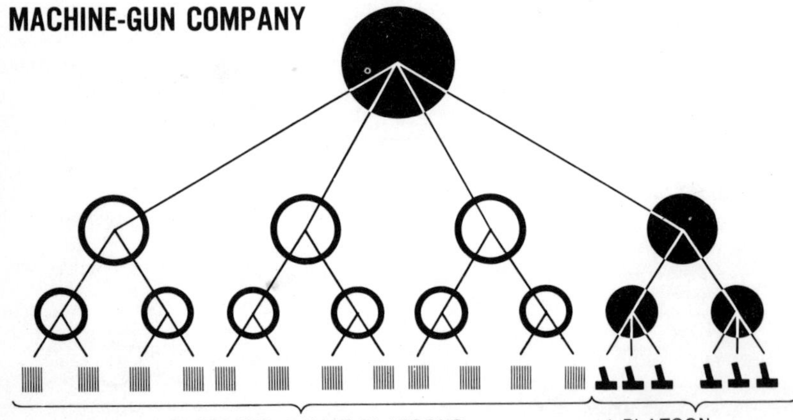

3. *Breakdown of German Infantry Regiment.*
4. *Breakdown of German Infantry Company.*
5. *Breakdown of German Machine Gun Company.*

Composed of three *Grenadier* battalions, the regiment included reconnaissance troops, engineers and signallers as well as infantry proper. Each battalion consisted of three infantry companies, one machine gun company, and a signal section.

Instead of a *Grenadier* battalion, some regiments possessed one *Jäger* battalion (literally "hunter battalion") which was trained as light infantry. Mountain troops, called *Gebirgsjäger*, were organised in a similar fashion to infantry regiments except that their heavy weapons—guns and machine guns—were carried by mules.

SMALL ARMS

Despite its rapid expansion between 1935 and 1940, equipping the *Wehrmacht* with small arms was no problem. Germany already had a flourishing home and export trade in sporting and commercial weapons produced by many firms of international repute. Rifles and pistols were needed for internal control and, as they were an inconspicuous form of rearmament, it was to be expected that the small arms industry would be well financed by Nazi money. Some firms owned by Jews were taken over by the Party, but they all retained their individual identities and enjoyed a good deal of freedom in research and development. Machinery and equipment were installed, research and development were guided on to officially sponsored projects, and plant was eventually converted from the production of commercial to military weapons.

As the process was gradual, however, it is not surprising that the German army was not fully equipped with standard weapons by 1939. Nevertheless, individual units had adequate supplies of equivalent arms of one pattern or another.

With full-scale rearmament and the declaration of war production came completely under official direction. Research and development were directed, but official specifications were not rigid, and firms were able to carry out development of production methods and a limited amount of experiment and design without official reference. Consequently, the designs that were submitted were dissimilar in detail, and this encouraged rivalry between firms, so giving a wider selection for the official choice. Those firms which exerted most influence on design were Walther at Zella-Mehlis; Haenel, Gustloff-Werke, Sempert and Krieghoff at Suhl; Erma-Werke at Erfurt; Mauser at Oberndorf/Neckar; Grossfuss at Döbeln; and Anker-Werke at Bielefeld.

Production was carried out by a great number of factories, many located in Thuringia and around Berlin, while factories in countries which had been overrun—such as Steyr in Austria, Beretta in Italy, and Fabrique Nationale in Belgium—were taken over. It is relevant to add that extensive use was made of plant taken over in Poland—a source of supply which was cut off later in the war, with disastrous effects.

SMALL ARMS AMMUNITION

Germany's pre-war rearmament programme was based on 9mm calibre ammunition for pistols and sub-machine guns, and 7.92mm calibre for rifles and machine-guns. Once the programme had started the urgency of the situation would not permit serious consideration of any alternative calibre. Nevertheless by 1939 ballistic research had led the Germans to two important conclusions: first, that 7mm was the most suitable calibre for accurate fire from a sniper's rifle; second, that small arms ammunition with a ballistic performance somewhere between the 9mm pistol and 7·92mm rifle ammunition was adequate for normal infantry battle ranges. If such ammunition were introduced, it could be expected that the designers would be able to produce lighter weapons for the infantry. With war clouds looming, however, the introduction of 7mm ammunition involved too great a change in production machinery and the idea was shelved. However, a medium performance ammunition was developed, and in 1941 the design of a bullet with the same calibre as the existing rifle ammunition but with a shortened version of the cartridge case was finalised. Existing machinery required only minor modifications to produce this new ammunition and by 1943, when a suitable new weapon had been introduced into the service, it was being produced in large quantities.

Known as the 7·92mm Pist Patr 43, the "intermediate" ammunition was the only outstanding development. Apart from this the story of German small arms ammunition does not embody any revolutionary change. The Germans exhibited their usual metallurgical skill of course, but shortage of materials exercised the strongest influence and was almost entirely the reason for each new development.

PISTOLS

Of the commercial output in Germany before the war, 7·65mm pistols were most common and 9mm pistols were not so numerous. With the outbreak of war the industry had not been converted to produce sufficient numbers of the army pattern 9mm "08" Luger to equip the forces, and issues were made up with other patterns of 9mm calibre. Officers were also able to buy "commercial" pistols. The S.S. and police were equipped with 7·65mm calibre pistols. It was also decided to supplement the Luger by introducing a 9mm pistol with a double action, which had been designed by Walther in 1938. As the 9mm Walther P 38 it was first issued in May 1940. From then on, this pistol started to replace the "08" as the standard weapon. Nevertheless, many firms continued to manufacture their own pattern pistols in addition to manufacturing the P 38 or its components.

So far as signal pistols were concerned, the standard pre-war weapon was the 26mm smooth-bore *Leuchtpistole*. Manufactured originally by Walther, it was produced between 1939 and 1941 by other firms. As the *Leuchtpistole* was made of a light alloy it is not surprising that economy in material necessitated changes in its specification, and in 1942 the *Leuchtpistole 32* appeared. This was very similar in design to its predecessor but was made of mild-steel pressings, gas-welded in assembly, and galvanised against corrosion. The one outstanding difference lay not with the actual pistol but with its ammunition since, in addition to the normal signal cartridges, the *Leuchtpistole* fired grenades.

The 26mm rifled-bore *Kampfpistole*—another Walther product— became a general army issue at the beginning of 1943. Except for its rifled barrel this weapon was very similar to the signal pistol, and it fired a range of pre-rifled cartridges and grenades. Because its offensive ammunition was soon outclassed by other weapons, however, the *Kampfpistole* became outmoded and was not used very extensively after 1943.

SUB-MACHINE GUNS

As with all the other small arms, the Germans had not produced enough of their standard army pattern sub-machine gun when war started. The result was that the troops were equipped with different patterns of German "machine-pistols". The official models were the MP 18 (Bergmann) and MP 38 (Schmeisser); but MP 28s (Schmeissers), MP 34s (Bergmanns), MP 34s (Steyr–Solothurns and Ermas) were also used.

The programme of replacing these unofficial weapons continued steadily as the MP 38 output increased. To simplify production— and so speed the process—the MP 38 was modified. In its new form, it appeared late in 1940 as the MP 40. This weapon, with the MP 38, had been produced in sufficient numbers to replace the unofficial weapons and to enable the MP 18 to be withdrawn from service by 1942.

Demands for, and subsequent losses in, the African Campaign caused a serious shortage of sub-machine guns. This led to the adoption of the Italian Beretta MP 38a and a modified version of it, the Beretta MP 38/42. The MP 40 also underwent a further modification, being fitted with a wooden butt and appearing as the MP 41. However, these three weapons were used on a limited scale and were soon eclipsed by the introduction of a completely new gun.

It was to be expected that the development of the short-case 7·92mm ammunition already mentioned would encourage the development of a suitable weapon, and in 1941, specifications for a new machine carbine to fire this ammunition were sent to the design departments of Haenel and Walther, who were also supplied with details and supplies of the ammunition. The Haenel model, the MKb 42(H), was completed in January 1942, and the Walther model, the MKb 42(W), about a month later. A limited number of both weapons were produced and underwent proving and troop trials. They were also exhibited to Hitler, who refused to countenance the introduction of a *Maschinenkarabiner* to his army—involving, as it did, the manufacture and issue of new ammunition. However, despite the official snub, development of the weapon continued. At first the general design of the MKb 42(H) was preferred, but as it fired from an open bolt—the MKb 42(W) fired from a closed bolt and was steadier in an unsupported firing position—it was decided to redesign it. Modifications were made and trials took place, and in 1943 the weapon was ready as the MP 43.

Hitler was again approached and asked to accept this new *Maschinenpistole* for army use. Persuaded perhaps by the urgent requirement for more sub-machine guns, Hitler now sanctioned its introduction and ordered a turn-over of production from the MP 40. In fact a considerable number of new weapons were produced in 1943 in two forms, known as the MP 43 and MP 43/1, without serious interruption of the MP 40 programme. The original MKb 42(H)s were recalled and modified, and the few existing MKb 42(W)s were issued to the army.

The next change came in 1943, when it was decided to cease production of the MP 43/1 and turn over the output entirely to the MP 43—which then became known as the MP 44. Later in the same year Hitler expressed his pleasure with the weapon by changing its name to *Sturmgewehr* 44—a "battle honour." As a result, it became known as the StuG 44. In the same year, a slide was added to the right side of the body of the weapon to accommodate a telescopic sight, the Zf 4, as used on the German army's snipers' rifles. The weapon was designed to take the standard rifle grenade discharger and it was intended to fire both explosive and armour-piercing grenades. But this aim was not achieved—probably because the already over-taxed ammunition industry was unable to produce a sufficient quantity of propelling cartridges.

It was at about this time, also, that the curved barrel attachment for the MP 44—the *Krummerlauf*—excited a good deal of allied interest. As it was an interesting development of an unorthodox principle, it is worth a minor digression. Experiments were carried out with curved barrels early in 1944. Reasonable results were obtained with a 30 degree curve, and it was hoped to achieve one of 90 degrees. Hitler was interested in the project, and the results achieved apparently justified his ordering 10,000 curved barrel attachments for the MP 44 in August 1944. These were to have a periscopic attachment easily fixed to the gun, and were intended for trench warfare. Some reached the army in 1945. A 90 degree curved barrel, which was to be mounted in a ball socket in the roof of tanks and armoured cars, never reached production stage.

The story of German sub-machine guns ended in the autumn of 1944, when the output of small arms had been seriously reduced by bombing and territorial loss. To maintain an adequate supply, it was decided to mass-produce a copy of the British "Sten" machine carbine, and so, on Hitler's instructions, 100,000 of them were ordered to be produced monthly. In fact nothing like this target was ever achieved. The original model, a very close copy of the "Sten", was referred to as the *Potsdam*, but after modifications, it went into production as the *Neumünster* or MP 3008.

RIFLES

At the outbreak of war the *Wehrmacht* was equipped with the "Type 98" rifles (*Karabiner* 98a, 98b and 98k), which fired standard 7·92mm ammunition, and whose design was based on the original pattern Gewehr 98 first designed by Mauser in 1898. The long 98a and its successor the 98b were both replaced gradually by the 98k— a shorter weapon—and the Gewehr 98/40, which was first introduced in October 1941. These remained the standard rifles of the German army to the end of the war. Mountain troops were equipped first with the Karabiner 98k but subsequently with the Gewehr 33/40, which was issued in November 1940.

Development of a 7·92mm self-loading rifle was started early in the war, and two separate designs, the Gewehr 41(W) and the Gewehr 41(M), were made by Walther and Mauser. Both were issued for troop trials in 1941, and as a result it was decided to produce the Walther model for general issue. As a standard weapon, this rifle first appeared in December 1942 as the Gewehr 41. Being rather heavy it underwent certain modifications, and the resultant new and lighter Gewehr 43 went into production early in 1943. Subsequently, its name was changed in 1944 to Karabiner 43.

The automatic rifle, not an army weapon, was peculiar to the parachute troops, who were part of the *Luftwaffe* organisation. Its production started in late 1942, and the early model was designed with many steel pressings. In 1943 it was slightly redesigned and, at the expense of weight, production time and cost, appeared as a more finished and robust weapon. Both versions were known as the *Fallschirmjägergewehr* or FG 42, and were produced in factories working under an air force contract. (At the time this weapon was considered an interesting hybrid, combining as it did the functions of a rifle with bayonet and a light machine gun with bipod.)

Telescopic sights were not manufactured for the German army before the war and, in the early months, therefore, only a few trained snipers were equipped with the Karabiner 98k fitted with a commercial rifle telescope, the ZF 39, manufactured by Hensoldt at Wetzlar. In 1940, however, an attempt was made to introduce an official rifle telescope, and the following patterns were produced in 1940 and 1941: ZF 40M for the Mauser S/L rifle; ZF 40W for the Walther S/L rifle; and ZF 41 for the rifle Karabiner 98k.

As demand far exceeded supply, privately owned hunting-rifle telescopes were called in for adaptation, and in 1942 the official pattern was simplified to speed production. These models were known as ZF 42 for the rifle Karabiner 98k; ZF 42M for the Mauser S/L rifles; and ZF 42W for the Walther S/L rifle. Then in 1943 another sight was produced for the semi-automatic rifle Gewehr 41, known as ZF 43B.

In the same year a telescopic sight, which became known as ZF 42, was produced for the early-model parachutist's automatic rifle, the FG 42. As might have been expected, all this multiplicity of design hindered output, and the trade was unable to cope with demand. It was not in fact until 1944 that a common design of telescopic sight, which could be manufactured easily and in which calibration was the only variation between models for different weapons, was finalised. The type of mounting, which had varied according to the weapon, was standardised by 1943. The sight was generally known as the ZF 4 and was fitted to the rifle Karabiner 98k, the self-loading rifle Karabiner 43, the parachutist's automatic rifle FG 42 (new pattern) and the machine-pistol MP 44.

Until about mid-1941 the firing of grenades from a rifle was not seriously contemplated. The first attempt was from a hollow spigot attached to the rifle muzzle by means of the bayonet fixture. A hollow-charge fin-stabilised grenade GG/P 40, was fired from this spigot using a propelling cartridge Patron G. This attachment appeared early in 1942 and was used in the African campaign, but not subsequently. In May 1942 a 3cm rifled discharger-cup appeared as army equipment. Called the Schiessbecher, it was used to fire spin-stabilised personnel and anti-tank rifle grenades. As an attachment for both the rifle and the machine-carbine MP 44 it remained in service until the end of the war.

Many attempts were made to design efficient silencers for German small arms, but the type which eventually appeared in 1944 was a close copy of the Russian rifle silencer. Known as the *Schalldämpfer*, it was issued on a limited scale to snipers for attachment to the rifle Karabiner 98k. Special low-velocity ammunition (*Gewehr-Nahpatrone*) had to be used with it.

In the autumn of 1944 Hitler decided to equip his "Home Guard" with small arms. As the production of standard weapons was barely sufficient to meet army requirement, however, it was obvious that what was needed was a simple weapon which could be turned out quickly in large numbers. The emphasis was on rifles and machine carbines, and in November 1944 pilot-models of various types of rifles were demonstrated to Hitler.

The Führer rejected all the single-loading rifles, chose one of the magazine rifles produced by the Deutsche Industrie Werke A.G. of Berlin, and subsequently ordered that a rifle firing the short-cased 7·92mm A.A.A. with a 10-round magazine was the objective. (A semi-automatic rifle which was demonstrated was rejected on the grounds that it was no quicker or cheaper to produce than the existing MP 44.) However, the problem of meeting army requirements, combined with the mounting losses of manufacturing capacity, prevented any great progress and few *Volksturm* weapons were produced. Nevertheless, they deserve some notice as they showed considerable ingenuity in their simple design, and skill in the use of pressings to eliminate machining operations.

MACHINE GUNS

Before war broke out the Germans had been unable to equip their army with a standard pattern machine gun, and during the period of rearmament what was issued consisted of a variety of old patterns of European manufacture. These were the MG 13, MG 08/15 and MG 15 light machine guns, the SMG 08 medium gun of German manufacture and the ZB 26 light machine gun of Czechoslovak manufacture. The dual-purpose light and medium machine gun MG 34, which went into production about 1936, had been chosen as the ultimate equipment of the army. But it was not a weapon designed for speedy mass production, and the Germans had only partially equipped with it when war started. MG 34S and MG 34/41, two modified designs which had been introduced in an attempt to improve output, were not produced in large numbers and it was March 1941 before the troops had their full quota of MG 34s, when MG 13, MG 08/15, MG 15 and SMG 08 could be recalled. The Czech ZB 26 continued as the standard machine gun of the home defence units for some time after this date.

Meanwhile the development of a machine gun of a design which would require skill and fewer machining operations in production had started before the war. When this design came to fruition early in 1941 it was referred to as MG 39/41. Following troop trials the decision was made to accept it, and production started in 1942. By the end of the year the new weapon, renamed MG 42 by this time, had started to reach the troops in the field. But it was not until November 1943 that distribution appears to have justified the official introduction of the weapon as stock to replace MG 34, though despite its fame MG 42 never did completely replace it. Furthermore MG 34 remained the standard tank machine gun throughout the war, although after February 1943 only one factory continued to produce spare parts and about 5,000 complete guns a month.

Some minor modifications were later made to MG 42, notably to the cocking handle. (The modified weapon was then referred to as the MG 43, and in 1944 it was given a chromium-plated bore and bolt in an attempt to reduce wear.)

MG 34 and MG 42, apart from the role of LMG on the bipod and MMG on the tripod, were also used in a light AA role—singly on the tripod, and in twin and quadruple mountings.

As an emergency measure it was decided in August 1944 to make an AA mounting to hold four MG 17 aircraft machine-guns.

INFANTRY ANTI-TANK WEAPONS

At the beginning of the war the German infantry anti-tank weapon was the *Panzerbüchse PzB 38* anti-tank rifle. The principle of its operation was taken from its Polish counterpart, which made use of a cartridge case of much enlarged diameter (13mm) and length to fire an armour-piercing bullet of rifle calibre (7·92mm) size. PzB 38 was a single-shot weapon, in which ejection of the spent case was effected by barrel recoil. This, of course, entailed production complications and so, during the first two years of the war, PzB 38 was replaced by PzB 39. This was generally similar to PzB 38, but had a fixed barrel and hence manual extraction and ejection as well as manual feed.

Although the performance of these weapons was very much superior to the British Boys anti-tank rifle, the outbreak of the war with Russia and the appearance of the T34 tanks rendered them inadequate. The first replacements for them simply took the form of an enlarged anti-tank rifle—*schwere Panzerbüchse 41 sPzB 41*. The PzB 41 was important because it employed the Gerlich principle of using a tapered barrel to "squeeze" a skirted projectile to a lesser calibre, thus resulting in increased muzzle velocity and hence in penetrative power. The increase in the size of the weapon entailed the use of a carriage and, although the complete equipment was kept remarkably light, its weight still made it unsuitable as the standard infantry weapon; and its penetration performance was not adequate to deal with the Russian tanks. During the course of development of this weapon, the possibilities of the adaptation to military use of the hollow-charge principle were realised by the Germans and the first hollow-charge projectile of German manufacture was used in the attack on Crete in 1941. This was a fin-stabilised grenade fired from a spigot discharger on the rifle. In 1941, a rifled discharger cup was introduced in place of the spigot and a series of hollow-charge rounds were introduced for the discharger cup.

The penetration of these hollow-charge grenades was of course superior to that of the anti-tank rifles: these were consequently withdrawn from service and converted, with the designation *Granatbüchse 39*, to take the discharger cup of the rifle. At the same time hollow-charge hand grenades and demolition charges were produced. The German General Staff specification for an infantry anti-tank weapon in 1941 was for a man-portable weapon to penetrate 200mm at 30 degrees at 150 to 200 metres—a specification never in fact attained by the Germans during the war. However, the discovery that rotation adversely affected the performance of hollow-charge projectiles led the Germans to seek some means of projecting an anti-tank missile without spin-stabilisation, whilst the discovery of the relation between the diameter of the hollow charge and its power of penetration convinced them of the need for a missile of comparatively large diameter. By 1942 a rocket weapon known as the *Püppchen*, which fired an 8·8cm projectile, had been produced. Known officially as *Raketenwerfer 43*, this weapon went into only limited production as its recoil was too much for a human firer to support. Although an experimental project, it was an important development because of its influence on later designs of the high/low pressure gun. Largely on the initiative of the notorious Colonel Dornberger, whose name is associated with the development of the V2, an open-breech weapon was developed. This was similar to the American bazooka and used the Püppchen's 8·8cm projectile. This German bazooka appeared in late 1943, officially designated *Raketenpanzerbüchse 54*, and soon became known popularly as both *Ofenrohr* (literally "stove-pipe") and *Panzerschreck* ("tank terror").

Meanwhile an entirely new weapon had been taken into service. Known at first as *Faustpatrone* ("fist cartridge"), its successor became famous in later models of the *Panzerfaust* ("armoured fist"). It is interesting to note that a Russian-made close-combat rocket used recently by suicide squads in Vietnam is a direct development of *Panzerfaust*. Known either as RPG-7 or as B-41, it is capable of penetrating 10 inches of armour at ranges of up to 500 yards.

GRENADES

In the preceding sections reference has been made to anti-tank grenades fired from spigot dischargers fitted to the infantryman's rifle. British and American understanding of the "grenade" is usually related to an anti-personnel hand-grenade, and it is unfortunate, though not altogether illogical, that the Germans referred to their mortars as *Granatwerfers* ("grenade throwers").

Hand-grenades were popular with the German infantry during the First World War, and although the standard grenade throughout the 1939–45 war was called the *Handgranate 39* (because of a few alterations in 1939), it was still the well-known stick grenade of 1914–1918. Most German infantrymen carried hand-grenades, and in every section of six men, two were designated as "bombers" and two more carried extra grenades.

INFANTRY MORTARS

Most Allied soldiers who fought against the Germans have an indelible impression of the effect of their mortars, and it is true that the Germans used these weapons with remarkable effect. Whether or not this impression is exaggerated, it is interesting to note that the German army formed the same impression of Russian mortars. The Germans often fired their mortars in batteries, which was not normally the Allied practice. They also used their mortars repeatedly, and with deadly effect, on positions from which they had just with-drawn and where Allied troops were in the process of digging in—such bombardments often being a prelude to a counter-attack. Furthermore, in North Africa the German 81mm mortar firing a 7lb bomb outranged the British equivalent 3in mortar (which fired a 10lb bomb) by 1,000 yards.

At the beginning of the war every German infantry company was equipped with three light mortars. One was the 5cm (2in) *Granatwerfer 36*. At the same time the machine-gun company of the infantry battalion had (in addition to its three platoons or four medium machine-guns) one platoon equipped with six 8cm (3in) *Granatwerfer 34* mortars. The latter remained in service throughout the war, but the 5cm mortar was, for its size, a costly and compli-cated weapon, and by 1942 it was already obsolete. In this year a new weapon was brought out to replace it as the company mortar—a short-barrelled lighter version of the 8cm mortar, known as the 8cm *Granatwerfer Kurz* (Short) *48* and commonly called *Stummel-werfer*. This weapon was issued on only a very small scale before being abandoned. The principal cause of this was the introduction in the same year of a 120mm (4·7in) mortar, which not only com-peted with the short 8cm for production facilities, but also by its issue began to displace the long 8cm in the infantry establishment.

HEAVY SUPPORT WEAPONS

Although such weapons cannot properly be considered infantry equipment, some mention must be made of the heavy support weapons included in the organisation of the German Infantry Regiment.

The Infantry Gun Company was a development partly from the artillery batteries assigned to the German infantry in the later stages of the First World War and partly from the trench mortar platoon, which was part of the regiment in that war. Originally equipped with mortars, in 1938 it was re-equipped with two different types of artillery pieces: the 7·5cm (3in) field gun and the 15cm (6in) howitzer. Similarly, the Infantry Anti-Tank Company was also equipped with artillery weapons: originally, 12 3·7cm anti-tank guns and four 2cm anti-aircraft guns. As the war progressed, however, there was a tendency to re-equip the Gun Company with the more cheaply produced and effective 120mm mortar *Granatwerfer 42* (GrW 42). This was openly copied from the Russian 12cm mortar, whose performance and mobility had made a marked impression on the Germans.

BAYONETS, FIGHTING KNIVES AND DAGGERS

The standard service issue bayonet was the Mauser Gew 98, with a 10in steel blade and wood or plastic grips. However, a large and extremely varied number of edged weapons were carried by the German forces in World War II for both combat and ceremonial use. Their study is of particular specialist interest and a separate companion volume will be included in this series.

WEAPONS ASSOCIATED WITH THE GERMAN INFANTRY

Finally, some mention must be made of two types of weapons which are sometimes erroneously thought of as German World War II infantry equipment.

Mortars used by the *Nebeltruppen* ("smoke troops") of the German army have been the cause of two misconceptions. First, such weapons were not included in the basic armoury of the Infantry Regiment; and second, their role was not an infantry one. The *Nebeltruppen* were formed not only for smoke laying but also, if not primarily, for chemical warfare. Since the decision to engage in chemical warfare was never taken, and since the laying of smoke was a comparatively infrequent occurrence, by far the greater number of rounds fired by the *Nebeltruppen* were high explosive bombs supporting infantry actions. (At the outbreak of the war, the *Nebeltruppen* were armed with a 10cm *Nebelwerfer* ("smoke projector") *35*. (*NbW* 35). When these weapons were withdrawn in 1942 the replacement was not the 12cm mortar, however. Because the primary role of chemical warfare had not been forgotten, the 15cm rocket launcher was deemed more suitable. This weapon accordingly, inherited the name *Nebelwerfer* and hence, by a mistaken association of ideas, the word *Nebelwerfer* is sometimes thought to be peculiar to rocket-launching weapons.)

Flame-throwers, used in conjunction with assault troops, were part of the armoury of the German engineers.

SMALL ARMS AMMUNITION

(German small arms ammunition manufacturers' code markings are listed in Appendix 5.)

Ammunition for pistols and sub-machine guns

Before the war the standard ammunition for pistols and *Maschinenpistolen* was the *9mm Pist Patr 08*, which had a brass cartridge, a brass cap and a lead bullet with a gilding metal-clad steel envelope. As early as November 1939, however, brass cartridge cases were being replaced by steel.

Shortage of lead caused the next development, and in February 1941 a new round—the *9mm Pist Patr 08(mE)*, in which the bullet was of mild steel with a lead cup in the base and a steel envelope—was issued. This ammunition had a marginally higher muzzle velocity than the *Pist Patr 08*. Further economies produced the *9mm Pist Patr 08(SE)*, which could be manufactured more easily than its predecessors. Here the bullet was made of solid sintered iron and the cap of zinc-plated steel. Production started in 1943, but it was not in common use until 1944.

Rifle Ammunition

In 1939 there were five standard types of infantry ammunition in service with the *Wehrmacht*. Differences were indicated by rings of different colours at the base of the cartridge and by markings at the point of the bullet.

Type	*Markings*
7·92mm Patrone (schwere Spitzgeschoss) (standard ball)—a lead core bullet with (gilding metal clad) steel envelope.	Green ring
7·92mm (Spitz mit Kern) (armour-piercing)—a steel core bullet with lead sleeve and a (gilding metal clad) steel envelope.	Red ring
7.92mm Patrone (Spitz mit Kern Leucht Spur) (armour-piercing with tracer)—a short, steel core bullet with a tracer pellet at the rear, with lead sleeve and a (gilding metal clad) steel envelope.	Red ring and blackened point
7·92mm B-Patrone (range observing)—a (gilding metal clad) steel envelope with white phosphorus in the nose, sealed by a lead plug, behind which is a detonator and striker in a lead sleeve.	Black ring, point blackened or chromed
7·92mm Patrone (Spitz mit Kern (H)) (armour-piercing)—a tungsten carbide core bullet with lead sleeve, and a (gilding metal clad) steel envelope.	Red ring, black bullet

In September 1940 the *7·92mm Patrone (S.m.E.)*—a mild steel core bullet with a lead sleeve, and a gilding metal-clad steel envelope—replaced the heavy pointed Patrone s.S. as the standard ball ammunition. Marking was a blue ring.

To achieve economy in the use of lead, the *7·92mm Patrone (S.m.E.lang)* —a mild steel core bullet with a lead ring, and a gilding metal-clad steel envelope—was developed and was introduced in 1944. In the same year a special ammunition known as the *7·92mm Gewehr–Nahpatrone* was introduced for use with the new rifle silencer. This was simply the old standard ball ammunition—with a lead cored bullet and reduced propellant charge to lower its muzzle velocity to 300 metres per second· Marking was a cartridge cap of zinc.

7·92mm Pist Patr 43

This short-case "intermediate" ammunition is the one really note-worthy German small arms ammunition development. A shortened version of the standard rifle steel cartridge case carried a boat-tailed 120 grains steel bullet (with the usual gilding metal cladding). A powder charge of 32 grains gave—when used in the MP44—a muzzle velocity of about 2250 feet per second.

Anti-tank Rifle Ammunition

Both the *Panzerbuchse 38* and its successor fired the same heavy pointed 7·92mm bullet which had a steel core and a tracer filling. Theoretically this ammunition was capable of penetrating 30mm of armour at a range of 100 metres (110 yards) or 25mm at 300 metres (330 yards). (cf. the penetrative performance of 7·92mm *Patrone S.m.K.*, which it was claimed would penetrate a 10mm steel plate at 100 metres.) This bullet also carried a tear gas capsule.

Special Performance Ammunition
Poison Bullets

"Poison" bullets for use in 9mm pistols, reputedly developed by the Germans, were said to be copies of Russian ammunition. The bullet had a normal jacket, with a cavity in the nose. Inside was a lead core, hollowed out. The base of the jacket contained a steel slug with tapered sides and a rounded cavity in the centre. Between the two, a glass ball capsule containing a cyanide solution was inserted. Impact caused the steel core to smash the glass against the lead core, pouring it through the nose cavity into the wound. (The only marking on this cartridge was said to be "K" on the base.)

Explosive Bullets

Early in 1945, the Germans were reported to be experimenting with a special type of "explosive" bullet. Lead azide, highly compressed, was used as the bullet core, which—like the poison bullets—had a nose cavity which was sealed with a thin metal disc. On impact the azide, inert at high pressure, became activated and detonated.

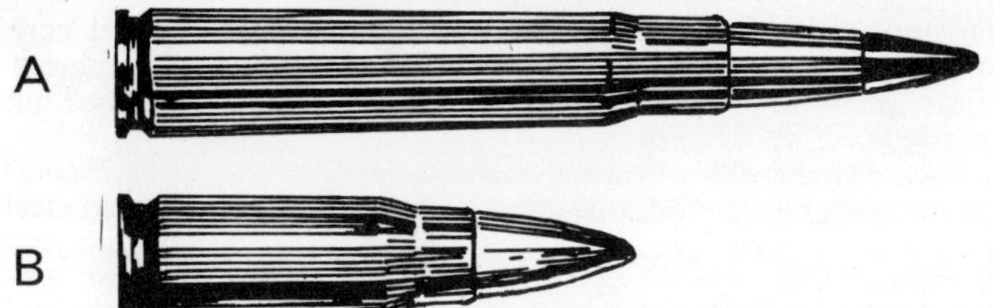

6. *German small arms ammunition (actual size).*
(A) 7·92 mm (STANDARD BALL)
(B) 7·92 mm (SHORT) PATRONE 43
(C) 9 mm (STANDARD BALL)

7. *Magazine of an MP 43 showing a round of 7·92mm "intermediate" ammunition.*

Much experimental work went into the development of a consumable cartridge case for all types of German small arms. Attempts to produce 7·92mm ammunition in which the empty case would collapse and form an integral part of the bullet when the weapon was fired were abandoned. So, too, was a development from this idea in which attempts were made to convert a 9mm bullet into a 9mm "rocket"—by making the bullet jacket integral with the cartridge and allowing the propellant gas to escape through gas vents in the side of the latter.

When hostilities ceased the firm of Polte at Magdeburg were developing an 8mm cartridge for use in a single-shot rifle. Utilising a plastic case of gelatinised nitro-cellulose and a conventional bullet, this development showed considerable promise.

9mm SELF-LOADING PISTOL 08 (LUGER)

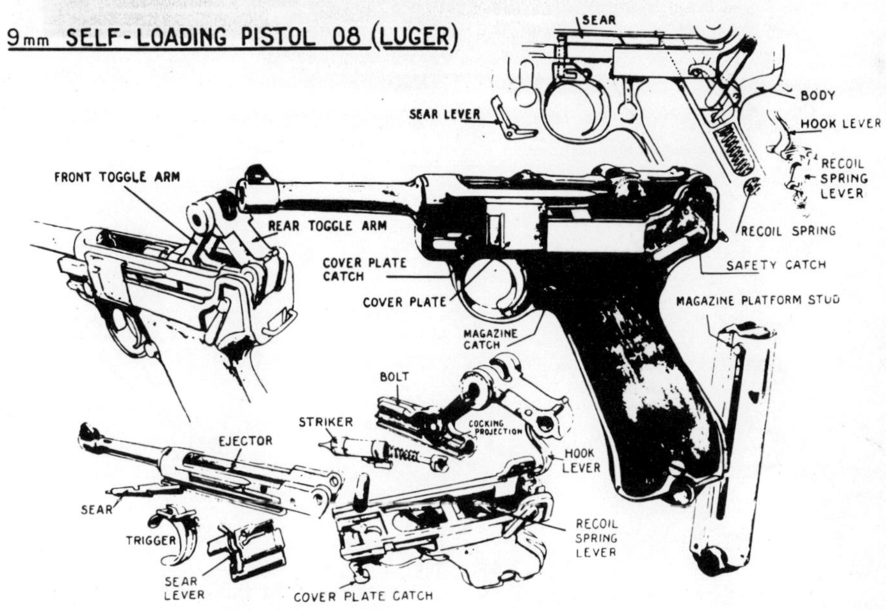

8. *9mm "08" Luger*

The Luger self-loading pistol, with its First World War and spy-story associations, is one of the best known pistols in the world. The original design by the American Borchardt was developed by a German named Lueger, and first manufactured under the name *Borchardt–Lueger*. Later this was corrupted to *Luger*—the name by which the pistol universally came to be known.

Basic Data of the Parabellum Luger

Parabellum is the military version. The 9mm Parabellum cartridge is recognised as one of greater power than is necessary off the battlefield. (The Treaty of Versailles prohibited German and Austrian arms firms from manufacturing pistols of the 9mm parabellum type. To overcome this the firms concerned made weapons in 7·65 and 7·63mm which could be converted to 9mm. Thus all parts, except the barrel and magazine, of a 7·65mm Luger can be interchanged with a 9mm Luger.)

Calibre: 9mm
Feed: Magazine
Capacity: 8 rounds
M.V.: 1,040–1,500 f/s
Bullet weight: 125 grains
Length overall: 8¾in
Length barrel: 4in
Weight: 30oz
Sights: Inverted "V" blade foresight; V notch rearsight, fixed
Range: Accurate to 75 yards
Range, Maximum: 1,200 yards
Method of operation: Recoil
Type of fire: Single shot only

9. *9mm Walther "P 38"*

The P 38, which replaced the Luger, has some unusual design features. Unlike the normal "automatic" the hammer of the P 38's double-action need not be kept fully cocked ready for action, and it can be safely carried in a loaded condition with the hammer lowered.

Basic Data

Calibre:	9mm
Feed:	Magazine
Capacity:	8 rounds
M.V.:	1,040–1,500 f/s
Length overall:	8½in
Length barrel:	4¾in
Weight:	34oz
Sights:	Fixed, Barleycorn foresight V notch back
Range:	Accurate to 75 yards
Method of operation:	Recoil
Type of fire:	Single shot (double action)

*Basic Data of some of the other German Pistols in service use between
1939 and 1945*
(Most of these weapons were privately purchased)

Walther 7·65 Mod. PPK S.L.

Calibre:	0·32in, 7·65mm
Feed:	Magazine
Capacity:	7 rounds
Overall length:	6in
Weight:	18½oz
Method of Operation:	Blowback
Type of fire:	Single shot (Double action)

Walther 3A 7·65mm

Calibre:	7·65mm
Feed:	Box Magazine, 7 rounds
Length overall:	6in
Weight:	18½oz
Method of Operation:	Blowback
Type of fire:	Single shot (Double action)

Sauer S.L. Pistol

Calibre:	7·65mm
Feed:	Magazine
Capacity:	8 rounds
M.V.:	Standard for this cartridge
Length overall:	6½in
Length barrel:	3⅜in
Weight:	22oz
Sights:	Fixed
Method of Operation:	Blowback
Type of fire:	Double action

Mauser 7·63mm

Calibre:	7·63mm
Feed:	Magazine
Capacity:	10 or 20 rounds
M.V.:	1,320–1,600f/s
Length overall:	12in
Weight:	45oz
Method of Operation:	Recoil
Type of fire:	Single and automatic

This First World War weapon was often issued in a wooden holster shaped like a rifle shoulder stock. When fitted to the holster it could be converted into a sub-machine-gun. As a pistol it was rather clumsy.

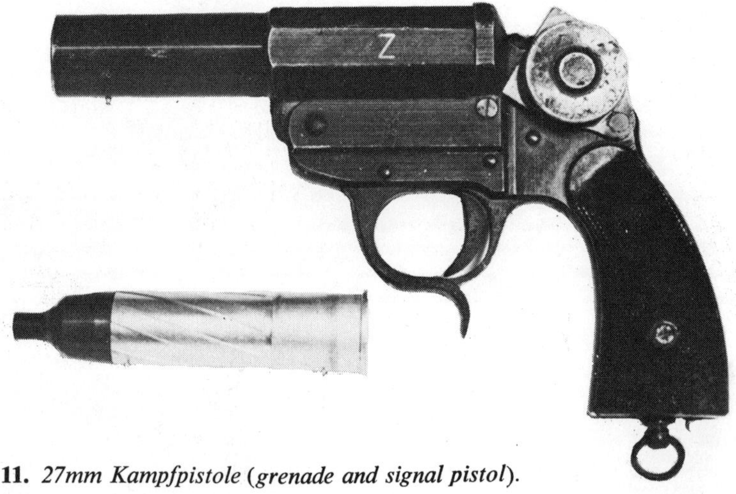

11. *27mm Kampfpistole (grenade and signal pistol).*

Signal Pistols

Illustrated (10) is a double-barrelled version of the 26mm Walther Leuchtpistole.

SUB-MACHINE GUNS

12. *MP 18 (Bergmann) 9mm*
Machine-Pistol

The Bergmann was one of the
earliest of the straight-blowback
action sub-machine guns. Like all
sub-machine guns—other than the
MP 44—it was designed to fire
standard 9mm parabellum "08"
pistol ammunition.

Basic Data

Calibre: 9mm
Feed: Box magazine, 32 rounds
    capacity
Overall length: 33in
Barrel length: 7¾in
Weight: 9lb
Sights: Fixed front; leaf rear
    50–1,000 metres
Range: Maximum accurate range
    approximately 300 yards. (A
    maximum range of 300 yards
    was claimed for the weapon,
    but this was an exaggerated
    performance estimate. A hun-
    dred to 200 yards is a more
    realistic figure.)
Method of operation: Blowback
Cyclic rate of fire: 540 shots per
    minute
Cooling: Air. (Heat is dissipated
    through a perforated casing
    around the barrel.)
Type of fire: Single shot or auto-
    matic. (A light pull on the
    trigger for single shot; pressing
    the trigger to its full extent
    activates the main trigger and
    gives automatic fire.)

Until 1934 the early model of the
MP 18 (MP 18.1) was one of the
two standard sub-machine guns of
the German paramilitary forces.
When the German general staff
decided to equip certain troops with
the *Maschinenpistole*, stocks of police
MP 18s were taken into army service.
It was not until 1943 that all first-
line units ceased to be equipped with
the MP 18.

13. *9mm MP 28* (*Schmeisser*) (sometimes known as the *Steyr-Solothurn*)

Developed by the German engineer Hugo Schmeisser from the MP 18, this weapon found almost universal favour before the war. Because of the limitations of the Versailles Treaty the manufacturing rights for Schmeisser's invention were transferred by the Haenel firm of Suhl to the Belgian firm of Pieper at Herstal. This also led to the adoption of the gun by the Belgian army as the *Mitraillette Model 34*. At the same time the Germans were manufacturing what was basically the same weapon in the Rheinmetall-owned Swiss factory at Solothurn—in order to evade the provisions of the Treaty. Furthermore, large quantities of the weapon were sold in the thirties to Japan and until they developed their own S.M.G. of similar design, the Japanese Imperial Army was almost wholly equipped with *Steyr-Solothurn MPs 28*.

Continued opposite

	Steyr-Solothurn	M.P.28
Calibre:	9mm	9mm
Feed:	Magazine, 30 cartridges	32 cartridges
M.V.:	1100–1600f/s	1100–1600f/s
Bullet weight:	125 grains	125 grains
Length overall:	32·25in	36·6in
Sights:	Leaf to 500 metres	Leaf 100 to 1000 metres
Weight:	10lb	9lb
Range:	Accurate, about 300 yards	Accurate 300 yards
Method of Operation:	Case projection (Blowback)	Blowback
Type of fire:	Single and automatic	Single and automatic
Cyclic rate of fire:	700rpm	600rpm
Cooling:	Air	Air

14. *MP 34 (Bergmann)*

The MP 34 was a development of the original MP 18, and its characteristics and performance are approximate to those given for its predecessor.

MP 34 (ERMA)

This 9mm *Machinenpistole* was basically the same design as the M 18, MP 28 and *Steyr-Solothurn* weapons. Manufactured in Germany by ERMA (Erfurt Maschinwerk), it was primarily for export. Like its contemporaries, it fired 9mm Parabellum "08" ammunition, and its basic data was very similar to what has been quoted for the *Steyr-Solothurn*. The one distinguishing feature about the ERMA was the provision of a forward hand grip, which was part of the furniture of the weapon and positioned about six inches in front of the trigger guard.

9mm MP 38

Developed by the Schmeisser factory at Suhl, this weapon was made entirely of metal and plastics, and because no wood was used it was considered to be of unorthodox design. A folding stock—enabling it to be used as a one-handed true "machine pistol" when folded or as a short-range automatic rifle when

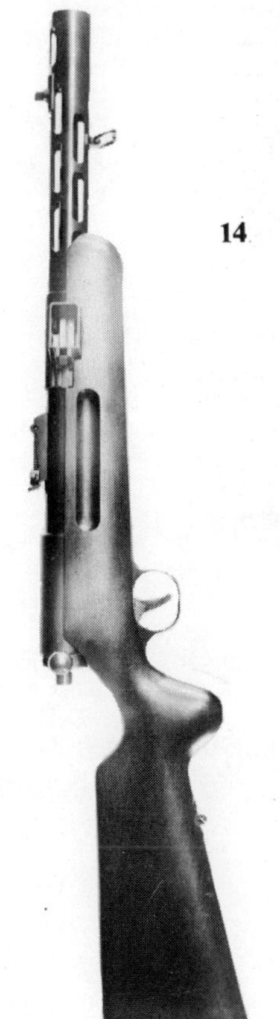

14

extended—was a breakaway from the conventional solid stocks of other machine carbines. The allied forces all looked upon the MP 38 (or its successor the MP 40) as being superior to their own equipment.

(For illustration and basic data, refer to the MP 40.)

15 and **16.** *MP 38 (Beretta)* and *MP 38/42*

The Beretta *Moschettõ* ("little musket") was the best of the Italian sub-machine guns—and the best known. The original model was taken into service by the Germans to augment their own stocks of *Maschinenpistolen*. The MP 38/42 was a modification (carried out on German instructions in the Italian factories) of the original Beretta. The difference is that the barrel jacket was dispensed with, the barrel was fluted to give a greater surface for cooling, and a Cutts-type compensator was added at the muzzle. Additionally the Italian-type sights were replaced by the Schmeisser type, which fold back.

Basic Data

    Calibre: 9mm
    Feed: Magazine, 10–20–40 rounds
    M.V.: 970 f/s
    Weight: 10·3lb with loaded 40 rounds magazine
    Sights: Leaf rear 100–500 metres; Fixed front
    Range: Accurate, about 300 yards
    Method of operation: Case projection
    Cyclic rate of fire: 400–500 rpm
    Type of fire: Single and automatic

15

16

17

17. *M.P.40* (sometimes known as the Schmeisser Parachutists' *Maschinen-pistole*)

The MP40, developed from the MP38, is one of the best known of the German infantry weapons. Originally designed as a parachutists' weapon, its popularity led to its supplanting all other types of sub-machine guns in the *Wehrmacht* and to its issue to tank crews as well as to ordinary infantry.

Basic Data

    Calibre: 9mm
    Feed: Magazine, 32 cartridges
    Length: With stock extended, 35in
    Length: With stock folded, 25in
    Weight: 9lb without magazine (loaded magazine weighed 23oz)
    Sights: Fixed front, standing rear leaf sighted to 100 metres and folding leaf sighted to 200 metres
    Accurate range: About 200 metres
    Method of Operation: Case projection (Blowback)
    Cooling: Air
    Cyclic rate of fire: 450–540rpm
    Type of fire: Automatic only

NOTE: A double-magazine version of the weapon was produced in 1943.

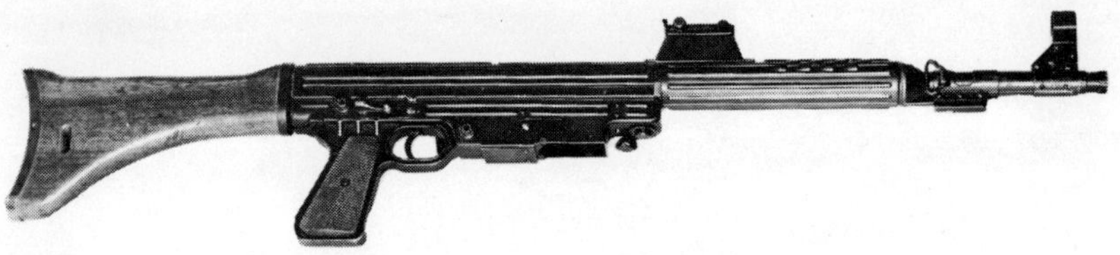

18

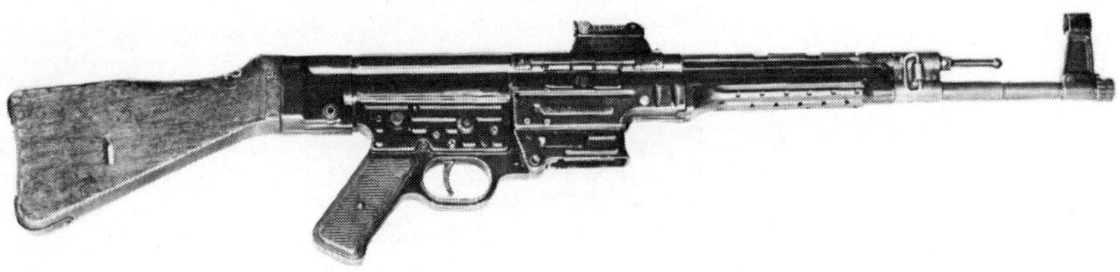

19

18 and **19**. *7·92mm MP 43 and MP 44 (Sturmgewehr 44)*

The MP 43 was the forerunner of the MP 44. As a sub-machine gun it was entirely out of the class of the *Maschinenpistolen* described so far. It was in fact an automatic carbine designed to fire the special shortened 7·92mm standard rifle cartridge. Its basic characteristics approximated to those of the MP 44, the production model. (The 1944 model carries the stamp "M.P.43" and also "K.44" (*Karabiner 44*).)

The MP 44 is noteworthy not merely for its efficient design but more especially for its simplicity of manufacture—being made up largely of cheap pressings. (The worth of its design was not in fact lost on the Russians, whose present automatic rifle embodies all the original characteristics of the MP 44.)

Basic Data

    Calibre: 7·92mm ("Intermediate" round)
Feed: Magazine, 30 rounds
M.V.: 2,250f/s
Bullet weight: 120 grains
Charger weight: 32 grains
Length overall: 36¾in
Length, barrel: 16in
Weight: 9½lb
Sights: Tangent and peep, 100 to 800 metres
Effective range: 300 yards
Method of operation: Gas
Type of fire: Single and automatic
Cooling: Air
Cyclic rate of fire: 800rpm

20 and **21.** *Krummerlauf (curved barrel attachment for the MP 44)*

Illustrated is the "Krummerlauf" attachment to the MP 44, designed not merely to fire round corners (a requirement arising from street fighting in Italy), but also for use in tanks. Russian infantry swarming over "closed-down" German tanks in battle were often found to be inaccessible targets to weapons fired through the pistol ports on the tanks. To dislodge them some sort of hose-pipe weapon was needed. (Similar circumstances reported by U.N. troops in Korea a decade later led to the experimental resurrection of this German innovation.)

It is clear that if the weapons were provided with a periscopic sight as shown the infantry could remain fully concealed whilst shooting from trenches and foxholes.

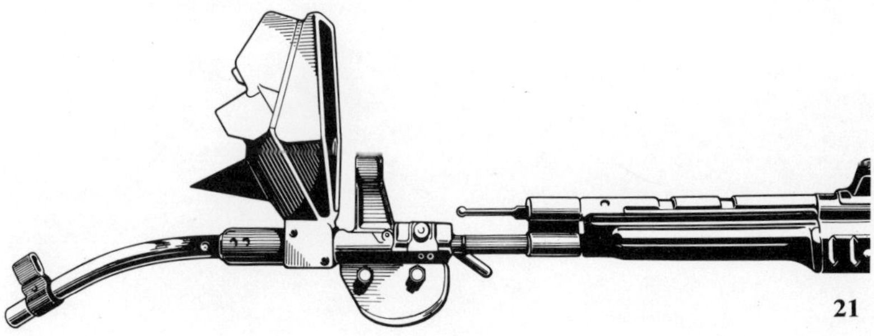

21

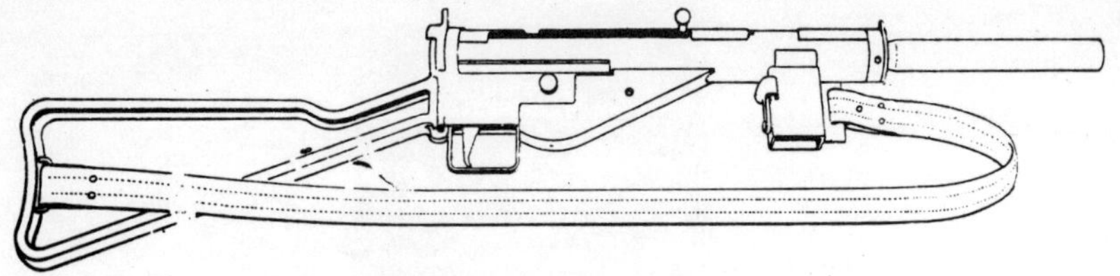

22. *9mm MP 3008* (*Neumünster*)

Hitler's "People's" resistance weapon—modelled on the British Sten-gun—was the MP 3008. Like the Sten, this weapon was intended as a cheap expendable mass-produced gun for home defence, and it was also noteworthy for its simplicity.

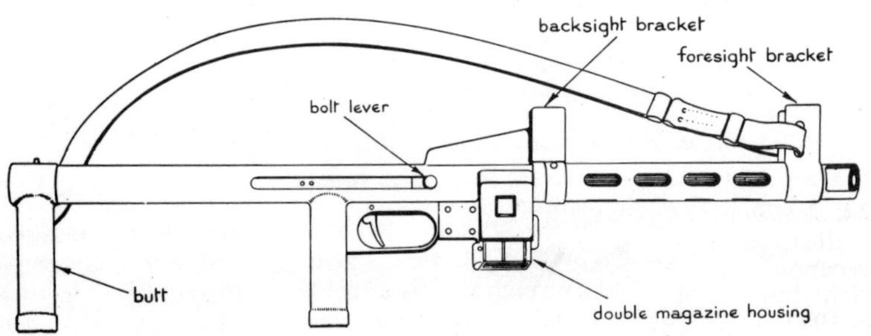

23. The EMP 44 which was also produced had similar characteristics but was, if anything, more crude even than the MP 3008.

24. *7·92mm Volkssturmgeschuss*
Illustrated is the "People's Weapon", a remarkable development by Haenal of Suhl. Designed in 1942, it was not issued until 1945. Like the British Sten—and its German copies—it was a simple cheaply-produced weapon. But as it used the 7·92mm "intermediate" ammunition it was in a superior class to the Sten and its counterparts.

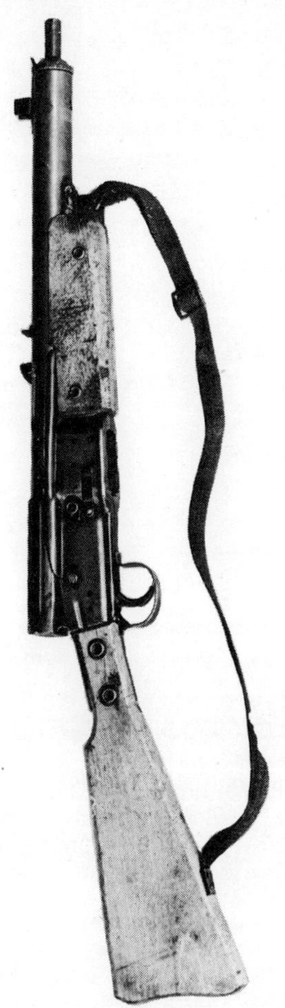

Basic Data

Calibre: 7·92mm ("Intermediate" ammunition)
M.V.: 2,250f/s
Feed: Magazine, capacity 30 rounds
Range: About 300 yards maximum
Barrel length: $14\frac{7}{8}$in
Overall length: $34\frac{7}{8}$in
Weight: $9\frac{1}{2}$lb
Sights: Fixed
Method of Operation: Blowback
Type of fire: Single shot

Mauser "Type 98" rifles and carbines

The Mauser Kar 98 series is world famous since Mauser designs were, until the general introduction of self-loading rifles, the most widely used for military purposes. Illustrated (**25**) is the 98a, but the basic data refers to the 98K. Appearance and data differ only in minor details, consequent mainly on the different barrel lengths.

25

Basic Data (Kar 98K)

Calibre: 7·92mm
Feed: Box magazine, 5 rounds capacity (no magazine cut-off)
M.V.: Approximately 2,800f/s with standard German rifle ammunition
Barrel length: 23·4in
Overall length: (without bayonet) 43·5in
Weight: 9 lb
Sights: V Blade front sight, rear leaf with open V notch graduated from 100 to 200 metres
Maximum range: Approximately 3,000 yards
Effective range: About 600 yards

Training Rifles

Between 1919 and 1939 all the great German arms firms (Mauser, Walther, Haenel and Genschow) manufactured air and 0·22 calibre rifles which, except for the barrel and breech arrangements, were exact copies of the Mauser 98a. This facilitated military training in violation of the Versailles Treaty.

26. *Mauser 7·92mm Gewehr 98/40*

The G 98/40 was the ultimate in the Mauser 98 series. First introduced in October 1941, it remained the German infantryman's standard weapon until the end of the war. (The *Gewehr 33/40* was similar in appearance, but shorter.)

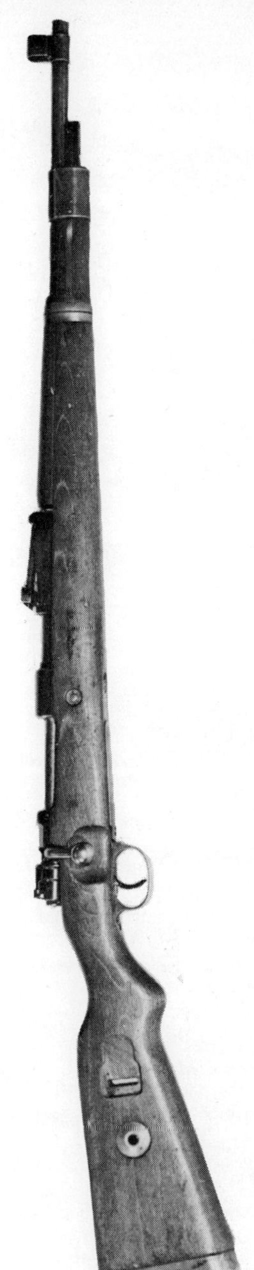

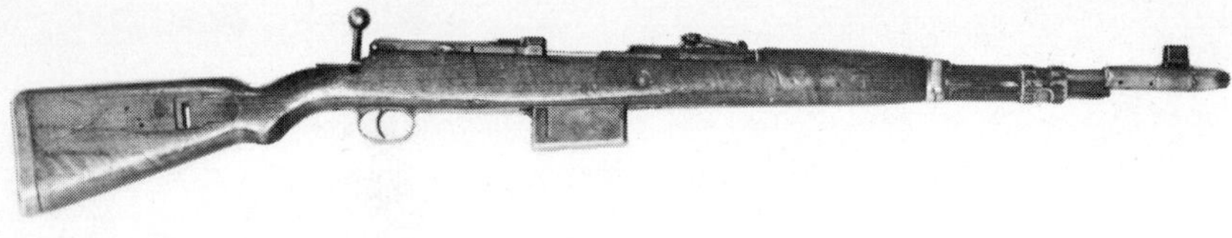

27

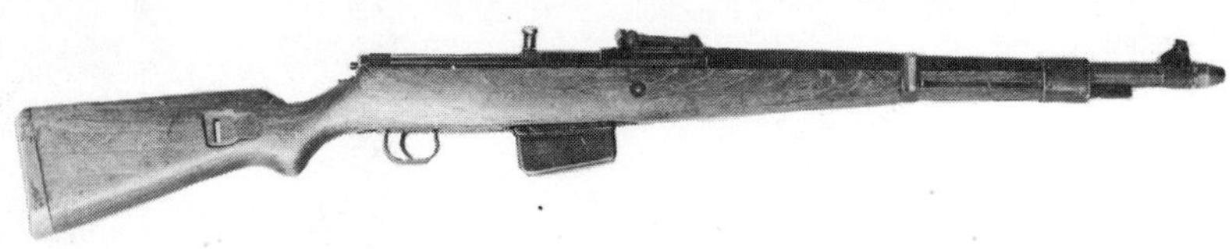

28

27 and **28.** *7·92mm Self-Loading Rifles G 41(M) and G 41(W)*

The similarity of design in both the Mauser and Walther models is apparent in the illustrations. Both used gas-operation, and they differed only in the G 41(W)'s having a longer piston and operating rod and no bolt-release button—as did the G 41(M). Both rifles were muzzle-heavy.

Basic Data

    Calibre: 7·92mm
    Feed: Magazine, 10 rounds (2 Mauser 5-round clips)
    M.V.: 2,882f/s
    Bullet weight: 154 grains
    Length overall: 45in
    Weight: 10 lb 14oz
    Sights: Leaf rear sight, 100–1,200 metres
    Range: (Effective) 600 yards
    Method of Operation: Gas
    Type of fire: Single shot, self-loaded
    Cooling: Air

29. *7·92mm Karabiner 43 (K 43)*

As this weapon was developed directly from the G 41(W), its basic data is only marginally different. However, paring the design down resulted in a saving of weight and a reduced overall length of half an inch. The weight was cut by approximately 2lb (to 8·9lb). But the basic saving in this weapon lay with the manufacturer. Everything that could be reduced from an engineering point of view was cut, and the number of parts that were machined was limited to bare essentials.

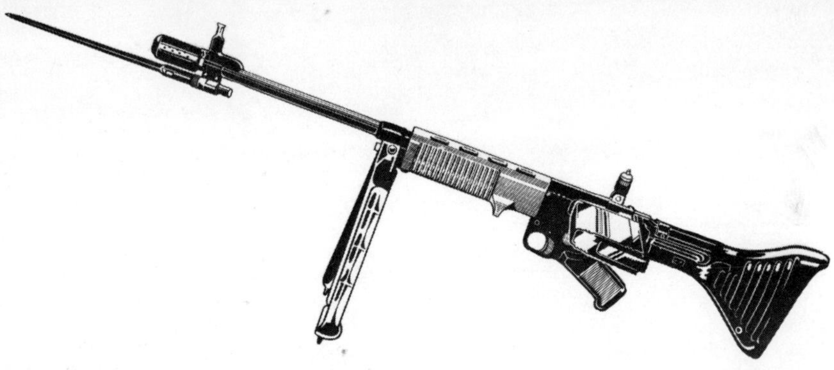

30

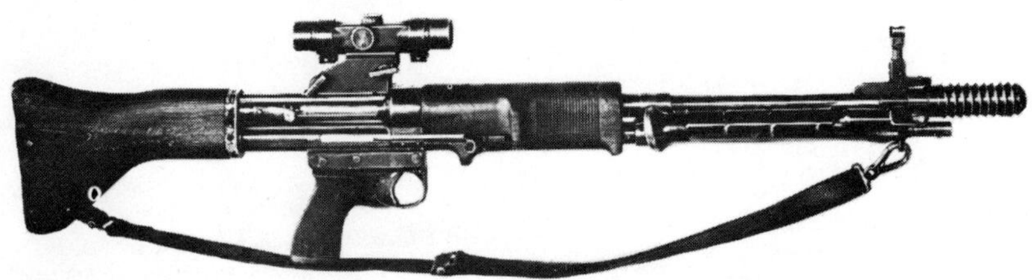

31

30 and **31.** *7·92mm Fallschirmjägergewehr 42 (FG 42)*

The FG 42 was a revolutionary small arms development. Designed originally as a dual-purpose weapon for parachute troops, it is a hybrid automatic rifle-cum-light machine gun-cum-sub-machine gun. Gas-operated, with a "straight through" action, it is fitted with a light folding bipod mount. When used in the role of a self-loading rifle the bolt operates from a "closed" position, thus giving maximum accuracy. When fired as a machine gun, however, the bolt remains open when the trigger is released, and this gives the maximum cooling effect necessary in automatic firing.

Basic Data

Calibre: 7·92mm
Feed: Box magazine, 20 rounds
MV: 2,250f/s
Length overall: 42in
Length barrel: 19in
Weight: 9¾lb
Sights: Front sight blade, back sight aperture 1–1,200 metres
Range: Effective: 300–600 yards
Method of Operation: Gas
Type of fire: Single and automatic

MACHINE GUNS
(*Light Machine guns*)

Until March 1941, when the MG 34 came into service, the German infantry was equipped with a variety of light machine guns. Even after that date some of these weapons lingered on with second-line units. Shortages even led to some garrison troops being equipped with other weapons which had formed the standard equipment of the armies of those countries Germany had overrun.

The Maxim 7·92mm MG 08/15 (not illustrated), an old water-cooled gun declared obsolete in 1936, was one of the first to be replaced. Like all Maxim guns it depended on a recoil operation and was locked by the well-known toggle-joint mechanism. Heavy (30lb on its bipod) and clumsy, its advantages lay in comparative reliability and the fact that it could—when mounted on a 51lb tripod—fulfil the role of a medium machine gun. Of the other guns the Dreyse MG 13 and the Czech ZB 26 fulfilled more realistically the light role required of an infantry section weapon. But the 7·92mm MG 15 (**32**) and 7·92mm Bergmann (**33**)—when fitted with bipod mounts—were both adequate. Both guns were used in armoured fighting vehicles, as illustrated.

32. *7·92mm MG 15*

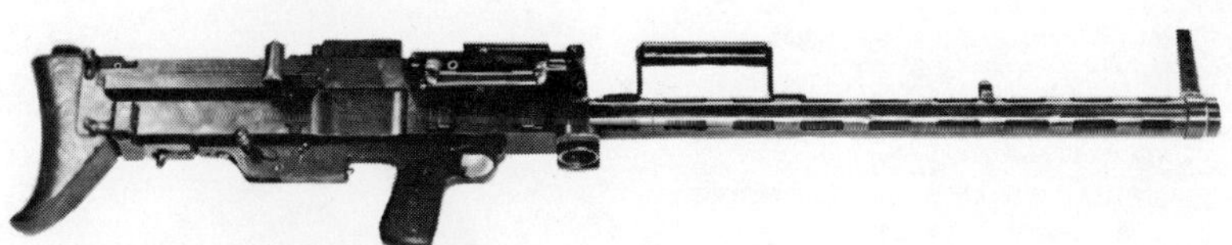

33. *7·92mm Bergmann MG*

34. *Czech 7.92mm ZB 26*

This weapon has the same origin
as the well known 0·303 inch
British Bren, which it closely re-
sembles in appearance and performance.

Basic Data

Calibre: 7·92mm
Feed: Magazine 30 rounds
M.V.: 2,624 f/s
Bullet weight: 154·3 grains
Length, overall: 3ft 11in
Length, barrel: 24in
Weight: 20lb
Sights: Barleycorn foresight, V
   backsight, Drum
Range: (effective) 600 yards
Method of operation: Gas
Type of fire: Single and automatic
Cooling: Air

44

35. *7·92mm Dreyse MG 13*

Manufactured by the famous Rheinmetall Company, this air-cooled light machine gun is distinguished by its folding butt and the long flash eliminator at the muzzle. A relic of the First World War, its appearance suggests the paternity of the MG 34 which was developed from it, and its characteristics are very similar thereto.

36. *7.92mm Madsen LMG*

The belt-fed Madsen was one of
the weapons taken into service in
1941 as an emergency measure, to
fill the gap between production and
the rapid expansion of the Wehrmacht. It was a clumsy weapon in
an LMG role and, as it fired 8mm
ammunition, supply problems made
it suitable for issue only to garrison
battalions.

Basic Data

    Calibre: 8mm
    Feed: Belt fed
    M.V.: 2,530 f/s
    Bullet weight: 196 grains
    Length, overall: 40 to 45in
    Length, barrel: 18¾ and 23⅛in
    Weight: 18lb
    Sights: Barleycorn front, Open V
      rear sighted to 1,000 metres
    Range: 600 yards
    Method of operation: Recoil
    Type of fire: Single and automatic
    Cooling: Air
    Rate of fire: 450rpm

(*Multi-Purpose Machine Guns*)
37 and **38.** *7.92mm MG 34 Light Machine Gun*

Because the Treaty of Versailles prohibited the Germans from manufacturing "heavy" machine guns (which in a modern context may be interpreted as medium guns), their designers set out to develop an all-purpose weapon which could be called a light machine gun but which could be adapted as a "heavy" weapon. Such thinking was in advance of its time.

The new weapon was developed at the Rheinmetall factory in Solothurn, in Switzerland, and the major problem of the overheating caused by sustained fire was overcome by providing for rapid changes of barrels. To cater for the high ammunition expenditure incurred with a sustained-fire weapon, a special metal link belt (**39** and **40**) was developed. Sections of this belt could be linked together by a single cartridge to produce an endless chain. Similarly, to counter the excessive vibration caused by the gun's firing long continuous bursts, a special mount was devised. The outcome was a weapon which could be used by one man as a section weapon or by a crew of three as a medium machine gun. It was a complete success and became the first really practical dual-purpose machine gun in the world.

Continued on page 49

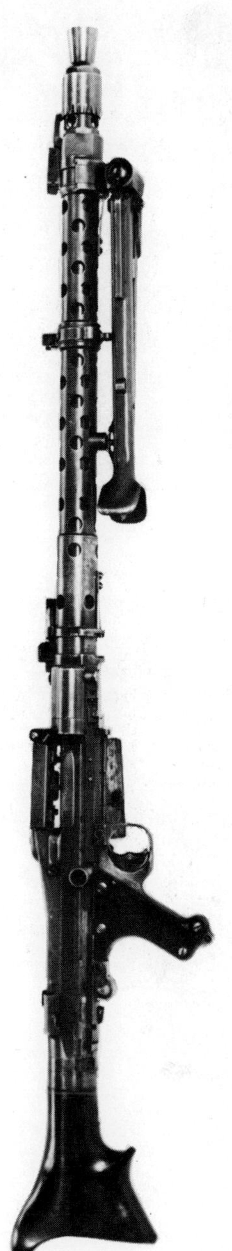

D

38. *7·92mm MG 34 Light Machine Gun*

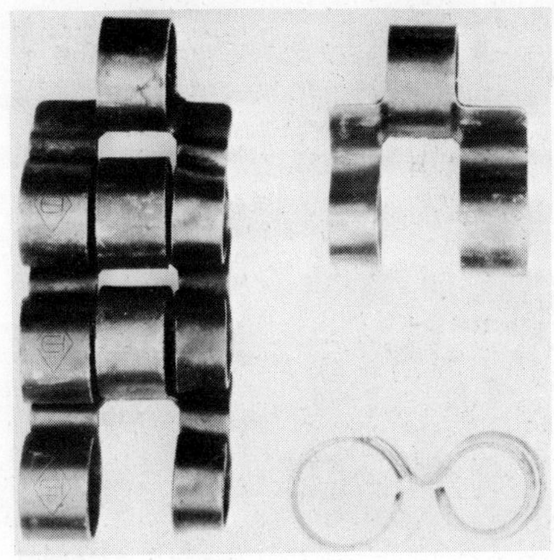

39

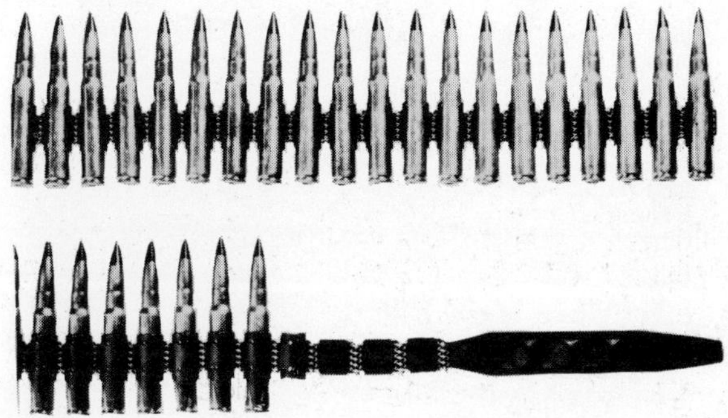

40

39 and **40.** *The special metal link belt for the MG 34*

7·92mm MG 34 Light Machine Gun: Basic Data

Calibre: 7·92mm
Feed: Metal belt, 50 rounds
M.V.: Vary from 2,500–3,000 f/s
Length overall: 48in
Length barrel: 23½in
Weight: 26½lb
Sights: V backsight, 200–2,000 metres
Range: Effective 2,000, on bipod 3,800, maximum 5,000 yards
Method of Operation: Recoil
Type of fire: Single and automatic
Cyclic rate of fire: 800 to 900rpm
Cooling: Air

41

41 and **42.** *7·92mm MG 42 Machine Gun*

Because of the cheapness, simplicity and ease of its manufacture, the MG 42 was one of the most remarkable light machine guns ever made. Like the MG 34, it was designed as a general purpose machine gun— capable of being adapted to the role of light, medium or anti-aircraft machine gun—and consequently suffered from the common defects invariably associated with all multi-purpose equipments. With a very high cyclic rate of fire, desirable enough in any anti-aircraft weapon, it was wasteful of ammunition when used in a ground role. In many ways this weapon typifies the efficiency of the German small arms industry. On the battlefield it gained fame and respect, and many British and American wartime soldiers will have vivid memories of the chilling rattle of its bursts of fire. However, it never completely superseded the MG 34.

42

Basic Data

Calibre: 7·92mm
Feed: Metal belt containing 50 rounds
M.V.: 2,500–3,000 f/s
Bullet weight: 154 grains
Length of gun: 4ft
Weight: 25lb with bipod
Sights: V backsight, 200–2,000 metres
Range: Effective range 2,000 yards, on bipod 3,800 yards; maximum range 5,800 yards (15,000ft in AA role)
N.B.—These ranges are those claimed by the Germans. By Allied standards the maximum effective range would be considered to be no more than 1,500 yards
Method of Operation: Recoil
Type of fire: Automatic only
Cyclic rate of fire: 900–1,200rpm
Cooling: Air

Like the MG 34, its predecessor, this gun is operated on the short recoil principle, assisted by a muzzle recoil booster. One unique feature of the gun is its novel locking system, in the two locking studs in the bolt head.

Plate 41 shows MG 42 in its LMG role, while it is shown fixed on its dual-purpose (ground/AA) mount for use as an MMG in plate 42.

43. *Breech block of the 7·92mm MG 42, showing the roller locking system*

44. *Periscopic aiming device*

This is for "under cover" firing of the MG 34 and MG 42 machine guns developed in 1944. This device was not a general service issue but is interesting if only to show the German fascination for small arms novelties.

45. *Panzerbüchse PzB 38* and *PzB 39 7·92mm Anti-Tank Rifles*

The design PzB 38 anti-tank rifle derived from the Polish M 38 (Model of 1938) rifle, although the bullet fired by the latter had only about half the penetrating power of the PzB 38. The PzB 39, which replaced the PzB 38, only had a marginally improved performance.

Basic Data

	PzB 38	PzB 39
Calibre:	7·92mm	7·92mm
Support:	Bipod	Bipod
Overall weight:	35lb	28lb
Length overall:	51in	50in
Length of barrel:	43in	43in
Method of Operation:	Single loaded	Single loaded
	Vertical type Breech block	Vertical breech block
		No automatic ejection
	Movable barrel	
	Automatic ejection	
Magazine capacity:	Single round loading	
Rate of fire:	10–12rpm	
Weight of cartridge:	2·97oz	
Length of cartridge:	4·5in	
Bullet:	pointed, hardened steel core, tracer filling and tear gas capsule	
Weight of bullet:	0·51oz	
M.V.:	3,970 f/s	

*Penetration in mm at 60° angle
    of impact (claimed)*

At 100 metres range:	30mm	
300 metres range:	25mm	
500 metres range:	—	

46. *7·92mm Granatbüchse 39*
(Grb 39)

As will be seen from the illustration, the Grb 39 was the old anti-tank rifle converted to take the standard rifle discharger cup (*Schiessbecher*). Primarily it was intended to fire one of the two standard hollow-charge anti-tank grenades (weights 8½ and 13½oz), but it could also be used to fire an anti-personnel grenade. Range was limited to 250 yards, and in practice the accurate range was probably much less.

54

47. *Raketenwerfer 43 (Püppchen)*

This interesting prototype weapon is shown on the modified wheeled carriage of the *Panzerbüchse 41* (an experimental enlarged anti-tank rifle intended as a replacement of the PzB 39 when the latter was found to be inadequate to deal with the Russian T 34 tanks). The *Püppchen* was pioneered by the firm of WASAG at Reinsdorf. Mounting it on a wheeled carriage, however, sealed its fate as an infantry weapon.

48. *Raketen Panzerbüchse* (*Panzerschreck*)

This anti-tank rocket launcher was copied from the design of a captured American bazooka. 3·5in calibre, and 5ft 4½in long, it weighed 20½lb (excluding a shield provided to protect the firer's face). Like its U.S .equivalent, the *Panzerschreck* fired a hollow-charge projectile—in this case, one weighing about 7lb. Its effective range was about 100 yards and the projectile is said to have been capable of penetrating 100 mm (about 4in) of armour

When fired, like all bazookas, the *Panzerschreck's* position was disclosed by a long flaming back-blast.

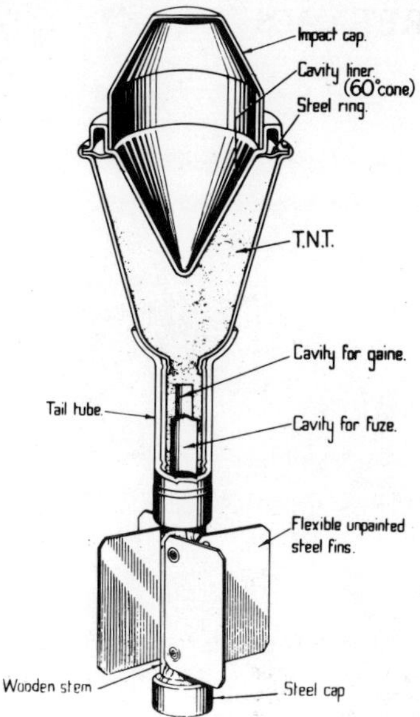

49 and **50.** *Faustpatrone* (*Panzerfaust*)

The *Panzerfaust* is probably the best remembered of German infantry anti-tank weapons. It served the same purpose as the *Panzerschreck* but, as it was nothing more than an oversized hollow-charge grenade, it could be carried by a single soldier in the forward infantry section. The original *Faustpatrone 30* was a dangerous weapon, distrusted by those to whom it was assigned. Later models were reasonably safe, however, and its effectiveness eventually resulted in its regaining popularity among the German infantry. Basically the weapon consisted of two parts: the projector—a short tube—and the projectile. When fired, a propellant charge in the tail of the projectile discharged it at a very low muzzle velocity. The weapon weighed 13¼lb and the projectile, which weighed 6¾lb, was said to be capable of blasting a hole through 8in of armour plate. The effective range was not considered to be greater than 100 yards.

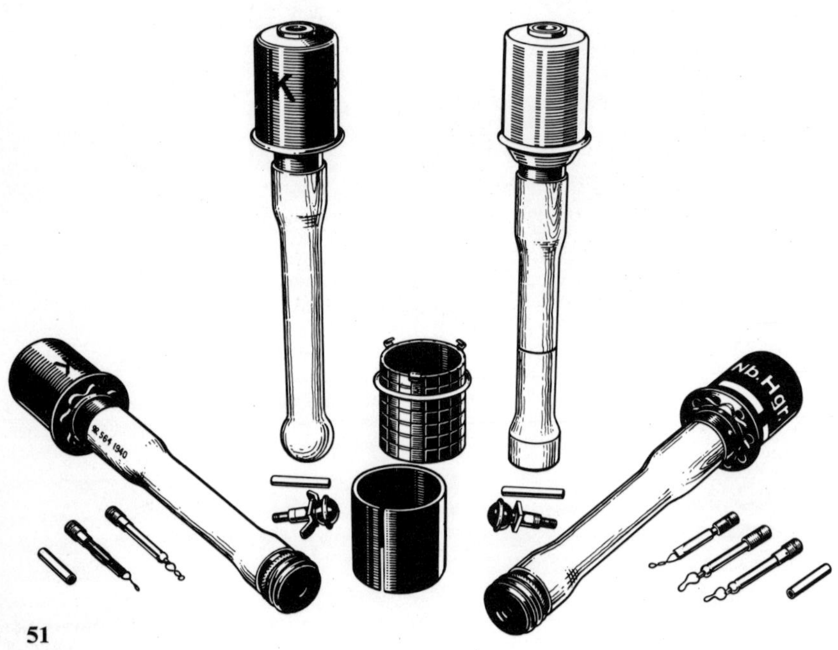

51

Grenades are among the oldest weapons of war, and the German stick grenade (**51**) was as famous in two world wars as the British Mills bomb and its 36 Grenade successor. The illustration shows its make-up. The hollow wooden handle screws into a socket in the body and holds the detonator. The complete grenade weighs about 1lb 13oz.

52. *5cm Granatwerfer 36 (5cm GrW 36)*

The smooth-bore, muzzle-loaded, trigger-fired GrW 36 was similar in many respects to the comparable British 2in platoon mortar, but only high-explosive bombs were produced for service use in the German weapon.

Basic Data

Calibre: 5cm (1·97in)
Weight in action: 14·0kg (30·8lb)
Barrel length: 465mm (19·3in)
Bore length: 350mm (13·8in)
Elevation: 750–1,600mils (42°–90°)
Traverse, total: 600mils (34°)
Bomb weight: 900 grams (31¾oz)
Charges: Primary only
M.V.: 75m/s (246fps)
Minimum range: 60 metres (65 yards)
Maximum range: 500 metres (547 yards)
50% zone at minimum range: 4 metres × 6 metres (4·4 × 6·6 yards)
50% zone at maximum range: 31 metres × 4 metres (39 × 4·4 yards)
Method of firing: Percussion
Sights: Not used
Method of transport: In horse limber; in two-man loads, or one-man load in emergency
Service ammunition: H.E. bomb Wgr 36

53. *8cm Granatwerfer 34 (8cm GrW 34)*

The GrW 34 was a conventional pattern gravity-fired mortar, similar to the British 3in or U.S. 81mm. H.E., H.E. "Rebound," smoke, and illuminating bombs were fired.

Basic Data

    Calibre: 81·4mm (3·2in)
    Weight in action: 56·7kg (125lb)
    Weight of barrel with breech piece: 18·3kg (40·3lb)
    Weight of bipod: 18·9kg (41.6lb)
    Weight of baseplate: 18·3kg (40·3lb)
    Barrel length: 1,143mm (45in)
    Bore length: 1,033mm (40·6in)
    Elevation: 700–1,600mils (40°–90°)
    Traverse, total: 160–260mils (9°–15°)—varies with elevation
    Bomb weight: 3·5kg (7·7lb)
    Charges: Primary cartridge and up to 4 secondaries: charges 1–5
    M.V.: Charge 1 75m/s (246fps)
           Charge 5 174m/s (571fps)
    Minimum range: 60 metres (65 yards)
    Maximum range: 2,400 metres (2,625 yards)
    50% zone at minimum range: 6 metres × 5 metres (6·6 × 5·5 yards)

50% zone at maximum range: 65 metres × 14 metres (71 × 15 yards)
Method of firing: Gravity
Sights: Dial sight RA 35
Method of transport: In vehicle or in three man-loads:
       (i) baseplate
      (ii) barrel with breech piece
     (iii) bipod with yoke and cradle
Service ammunition: 8cm Wgr 34   HE
                         8cm Wgr 34 Nb Smoke
                         8cm Wgr 38   HE-Rebound⎱ discontinued
                         8cm Wgr 39   HE-Rebound⎰ 1942/3
                         8cm Wgr 38 Deut.
                         Target indicating
                         8cm WLtGs 4447 Illuminating

Short 8cm Model 42—Kurzer 8cm Granatwerfer 42 (Kz 8cm GrW 42)

The Kz 8cm GrW 42 was of a similar pattern to the GrW 34 but shorter and simpler in design. It used the same ammunition and sights, but the barrel length was reduced from 45in to 29·4in. Consequently, the maximum range was reduced from 2,625 yards to 1,200 yards.

WEAPONS ASSOCIATED WITH THE GERMAN INFANTRY

Flamethrowers

Illustrated (**54**) is the man-pack flamethrower M 41. In design, this weapon closely resembled its British and U.S. counterparts, having a range of about 65 yards.

54

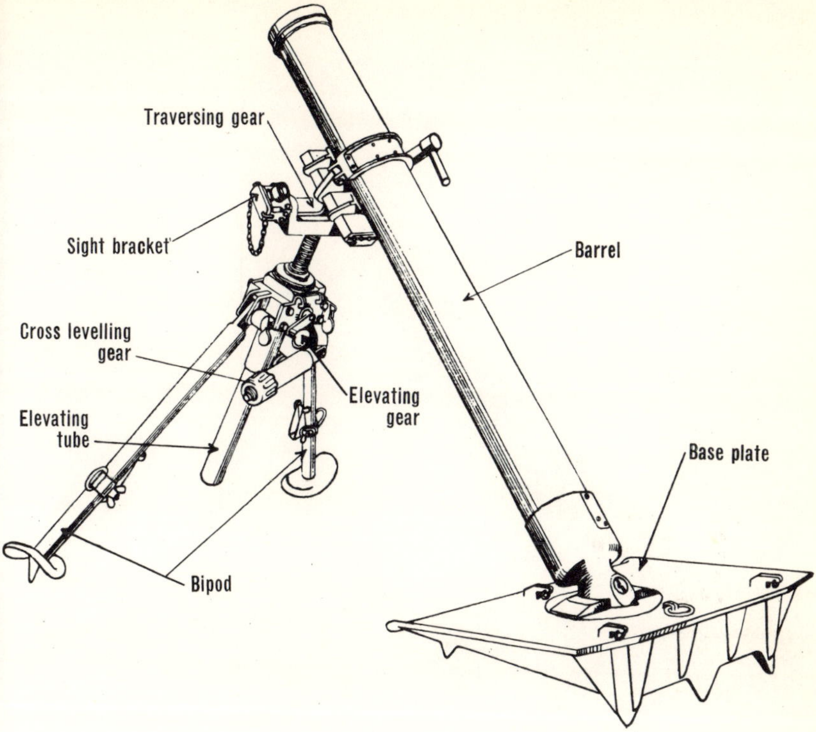

55. *10cm Smoke Mortar 35 (10cm Nebelwerfer 35)*

This was the standard smoke weapon of the German army. It could also fire high explosive ammunition, however, and was sometimes used by airborne troops as a close infantry support weapon. In design, it closely resembled the 8cm GrW 34.

APPENDIX 1—*Comparative Table of the Principal German, British and U.S. Small Arms of World War II*

Nationality	Weapon	Calibre	Weight	Approx. maximum range	Effective range	Cyclic rate of fire rpm
GERMAN	Pistol "08"	9mm (0·354in)	2lb	Up to 30 yards	—	—
	Walther P 38		2lb 5oz			
BRITISH	Pistol 0·380 No. 2	0·38in	1lb 11½oz	Up to 30 yards	—	—
AMERICAN	Pistol Colt 0·45 1911 Al Automatic	0·45in	2lb 11oz	Up to 30 yards	—	—
GERMAN	Rifle 98	7·92mm (0·311in)	9·05lb	2,200 yards	600 yards	—
	S/L Rifle G 41 W		10lb 14oz	2,200 yards	600 yards	—
	FG 42 Auto Rifle		9·75lb	1,500 yards	300–600 yards	—
BRITISH	SMLE Rifle No. 1	0·303in	8lb 10½oz	2,000 yards	600 yards	—
	Rifle No. 3 Mk 1 (Pattern 14)	0·303in	9lb 6oz	2,000 yards	600 yards	—
AMERICAN	S/L Garand	0·30in	9lb 7oz	2,000 yards	600 yards	—
	S/L Carbine M 1	0·30in	5lb 3oz	2,000 yards	300 yards	—
GERMAN	Schmeisser MP 40	9mm	9lb	200 yards	30 yards	450–540
	MP 44	7·92mm	10lb	200 yards	30 yards	800
BRITISH	Sten Mk III	9mm	6·5lb	200 yards	30 yards	500–550
AMERICAN	M 3 SMG	45in	6lb	200 yards	30 yards	450
GERMAN	MG 34	7·92mm	26lb	2,750 yards	800 yards	800–900
	MG 42		24lb	2,750 yards	800 yards	1,200
BRITISH	Bren	0·303in	23lb	2000 yards	800 yards	450–550
	Vickers		30lb *without* tripod	3,500–4,000 yards	800–1,200 yards	500
AMERICAN	Browning Auto Rifle M 1918	0·30in	17lb	2,000 yards	600 yards	500
	Johnson LMG (used by the U.S. Marines)		15lb	2,000 yards	600 yards	450–750
	Browning MMG		33lb	3,500–4,000 yards	800–1,200 yards	400–520

Comparative Table of German, British and

Nationality	Designation	Calibre inches	Total WT lbs	Remarks
1. *LIGHT MORTARS*				
German	5cm GrW 36	2	31	Muzzle loaded, smooth bore mortar. Broke down into two loads: the baseplate, traversing and cross-levelling gear; the barrel and elevating gear. Mortars made after 1938 had no sights. Mortar would only fire high angle
British	2in Mk 2* Mk 2** Mk 7	2	21	2in mortar with a large baseplate
	2in Mk 2*** Mk 7** Mk 8	2	11	2in mortar with full-length barrel and spade base-plate
	2in Mk 7*	2	9	AIRBORNE 2in mortar with shortened barrel and spade base plate
American	60mm M 2	2·3	42	A muzzle loading smooth bore weapon of conventional Stokes–Brandt design. This was a platoon weapon in the American army
	60mm Mortar T 18E6	2·3	19½	This weapon consisted of a 60mm M2 barrel fired from a spade baseplate— making it not unlike the latest British 2in mortar. Only a limited number was manufactured
2. *MEDIUM MORTARS*				
German	8cm short GrW 42	3·2	56	A cut down version of the 8cm long GW 34 Weight compared with the 8cm long GW 34 as follows; *Short* *Long* Barrel 22¼ 40 Bipod 21½ 40 Baseplate 10½ 44
German	8cm long GrW 34	3·2	125	A muzzle loaded, smooth bore mortar of conventional Stokes–Brandt design
British	3in Mk 5	3·2	112	The lightest British 3in mortar. Conventional Stokes–Brandt design
American	81mm M1	3·2	141	A muzzle loading smooth bore weapon of conventional Stokes–Brandt design. This weapon was an exact copy of the Italian 81mm mortar and corresponded to the 3in British mortar
American	81mm T27 (Short 81mm)	3·2	66	A cut down version of the 81mm M1 using a 60mm mortar baseplate. Only a limited number was made

U.S. Infantry Mortars of World War II

| Bombs | | No. of charges | Max. range yards | Remarks |
Type	WT lbs			
HE	2	1	347	This range was considered inadequate in N. Africa and it soon became obsolete
HE	$2\frac{1}{4}$	1	500	
Smoke bursting	$2\frac{1}{4}$	1	500	
Smoke emission	2	1	500	
Illuminating	1	1	—	Reached a height of 500ft at 85 degrees
Multi-star				(Max. range of the airborne mortar was 350 yards)
White	2	1	—	
Red	1	1	—	
Green	1	1	—	
Red and Green	1	1	—	
Smoke bursting	4	—	1,610	
HE	3	5	2,017	Max. practical range of the T 18E6 was 350 yards
Illuminating	4	5	1,075	
HE	$7\frac{3}{4}$	3	1,200	
Bursting smoke	$7\frac{3}{4}$	3	1,200	
Indicating	$7\frac{3}{4}$	3	1,200	Emitted blue smoke
Incendiary	—	—	—	
Illuminating	—	—	—	
Jumping bomb	$7\frac{3}{4}$	3	1,200	

All bombs were the same as for the long mortar but used a maximum of two secondaries instead of four

| Bombs | | No. of charges | Max. range yards | Remarks |
Type	WT lbs			
HE	$7\frac{3}{4}$	5	2 625	
Bursting smoke	$7\frac{3}{4}$	5	2,625	
Indicating	$7\frac{3}{4}$	5	2,625	Emitted blue smoke
Incendiary	—	—	—	
Illuminating	—	—	—	
Jumping	$7\frac{3}{4}$	5	2,625	
HE	10	2	2,790	
Smoke bursting	10	2	2,790	
Base ejection smoke	—	2	2,790	Fitted with a time fuse, it was also used for coloured smoke and flares (Only 5,000 were made)
Star	10	1	—	
Smoke emission	10	2	2,790	
HE	7	6	3,280	
HE	$10\frac{3}{4}$	4	2,560	
HE	15	4	1,280	
Smoke bursting	$11\frac{1}{2}$	4	2,470	
Illuminating	$11\frac{1}{2}$	4	2,200	
HE	7	—	1,935	
HE	$10\frac{3}{4}$	—	1,360	The ammunition was the same as for the 81mm M1, but the range was naturally less
HE	15	—	—	
Smoke bursting	$11\frac{1}{2}$	—	—	
Illuminating	$11\frac{1}{2}$	—	—	

APPENDIX 3

Details of German Infantry Mortar Ammunition in Service During World War II

Type and German abbreviation	Weight lbs	Weight & nature of filling	Colour	Number of charges	Weight Primary grains	Weight Secondaries grains	Overall length ins	Type of fuse	Muzzle Highest charge f/s	Velocity Lowest charge f/s	Maximum range for each charge yards	Remarks
5cm HE 36 (5cm Wgr 36)	2¼	4½oz TNT	Dull red	1	62		8·5	Nose percn. Whr Z 38	246	—	570	
8cm HE 34 (8cm Wgr 34)		18oz TNT	Chocolate brown									FS = Sulphur trioxide mixture
8cm Smoke 34 (8cm Wgr 34 Nb)		16oz sulphur trioxide	Dull red					Nose percn. Wgr Z 38 Wgr Z 38 St.			I— 591 II—1,094 III—1,597 IV—2,078 V—2,625	Fired from both 8cm.
8cm HE 38* (8cm Wgr 38 umg)	7¼	13½oz TNT	Green	5	154	139(4)	12·9		571	426		Mortar 34 and short 8cm Mortar 42.
8cm HE 39 (8cm Wgr 39 umg) 8 cm Indicator 38 (8cm Wgr 38 Deut)		13½oz TNT FS	Field grey Green, with "blau" overprinted in black	Charge I = Primary II = Primary + Secondary, etc.								*Converted to normal HE bombs from the earlier air-burst types.
10cm HE 35* (10cm Wgr 35 Spr.)	16	3¾lb TNT	Grey green	3	231	324(4)	17·1	Nose percn Wgr Z 33			III—3,300	Fired by 10cm Smoke mortar 35.
10cm Smoke 35 (10cm Wgr 36 Nb.)		3¾lb FS	Grey green	Charge I = Primary + 1 Secondary Charge II = Primary + 2 Secondaries Charge III = Primary + 4 Secondaries								*Previously known as Wgr 37.
10cm HE 40 (10cm Wgr 40 Spr.)	19	3¾lb TNT	Grey green	3	Charge I = 2·1oz Charge II = 5·2oz Charge III = 8·8oz			Nose percn. Wgr Z 33	1,017	427	III—6,780	Fired by 10cm BL Nebelwerfer 40
10cm Smoke 40		3¾lb	Grey green									
10cm Smoke (10cm Wgr 40 wKh Nebel)		3¾lb FS	Grey green									

German Small Arms Manufacturers' Codes

A.

aak — Waffenfabrik Brünn AG, Prague
ac   — Carl Walther, Zella Mehlis, Thuringia
aek — F. Dusek Waffenerzeugung, Opono bei Nachod
amn— Mauser Werke, Waldeck bei Kassel
ar   — Mauer Werke, Borsigwalde bei Berlin
asb — D.W.M. AG, Borsigwalde bei Berlin
auc — Mauser Werke AG Marien Str Köln-Ehrenfeld
awt — Württembergische Metallwarenfabrik AG, Geislingen
axs — A. Krupp, Berndorfer
ayf — "Erma" (Waffenfabrik), B. Geipel, GmbH, Erfurt
azg — Siemens-Schuckert Werke AG, Berlin

B.

bcd — Gustloff Werke, Weimar
be   — Berndorfer Metallwarenfabrik Arthur Krupp AG, Berndorf,
        Niederdonau
bh   — Brünner Waffenfabrik AG, Brunn (Brno, Czechoslovakia)
bjv — Böhmisch-Mährische Kolben AG, Danek, Prague
bkp — Gewehrfabrik H. Burgsmüller & Sohn, GmbH, Kreiensen/
        Harz
bkq — Röhrenwerk Johannes Surmann, GmbH, Arnsberg/W
bky — Böhmische Waffenfabrik AG in Prag-Werk, Ung-Brod.
        (Moravia)
bmv— Rheinmetall-Borsig AG, Sömmerda, Thuringia
bmz— Minerva Nähmaschinenfabrik AG, Boscowitz
bnd — MAN-AG, Nuremberg
bnz — Steyr-Daimler Puch AG, Werke Steyr, Steyr, Austria
bpr — Johannes Grossfuss Metall-u-Lackierwarenfabrik, Dobeln,
        Saxony
br   — Mathias Bäuerle Laufwerke GmbH, St. Georgen, Schwarz-
        wald

bvl — Th. Bergmann & Co. Abtlg-Automaten-u-Metallwaren-
      fabrikation, Hamburg
bwo — Rheinmetall-Borsig AG, Düsseldorf
bxb — Skodawerke, Pilsen
byf — Mauser Werke, Oberndorf
bym— Genossenschafts-Maschinenhaus der Büchsenmacher, Fer-
      lach, Austria
bzt — Wolf, Fritz, Rob S, Gewehrfabrik, Zella Mehlis, Thuringia

C.

ce — J.P. Sauer & Sohn Gewehrfabrik, Suhl, Saxony
ch — Fabrique Nationale d'Armes de Guerre, Herstal/Liège,
      Belgium
cof — Carl Eickhorn, Solingen
cdo — Th. Bergmann & Co., KG, Velten/Main
chd — Deutsche Industrie Werke AG, Berlin
con — Franz Stock Maschinen-u-Werkzeugfabrik, Berlin
cos — Merz Br, Frankfurt/Main
cpo — Rheinmetall-Borsig, Berlin/Marienfelde
cpq — Rheinmetall-Borsig, Guben
cpp — Rheinmetall-Borsig, Breslau
crs — Paul Weyersberg & Co., Solingen
cvl — WKC Waffenfabrik GmbH, Solingen
cyg — Spreewerk GmbH, Metallwarenfabrik, Berlin

D.

df b — Gustloff-Werke, Suhl
dot — Waffenwerke, Brünn AG, Brno, Czechoslovakia
dou — Waffenwerke Brünn AG, Bystrica, Czechoslovakia
dov — Waffenwerke Brünn AG, Vsetin, Czechoslovakia
dow — Opticotechna, formerly Waffenfabrik Brünn AG, Prerau,
      Czechoslovakia
dox — Waffenfabrik Brünn AG, Werk Podbrezova
dph — I.G.-Farbenindustrie AG, Werk Autogen, Frankfurt/Main
dsh — Waffenwerke Ing F. Janecek, Prague
dql — Remo Gewehrfabrik, Gebrüder Rempt, Suhl
duv — Berliner-Lübecker Maschinenfabriken, Lübeck
duw — Deutsche Röhrenwerke AG, Mühlheim-Ruhr

E.

egy — Ing. Fr. August Pfeffer, Oberlind, Thuringia

F.

fnh — Böhmische Waffenfabrik AG, Strakonitz, Prague
fue — Mechanische Werkstatt (Skoda), Dubnica
fwh — Norddeutsche Maschinenfabrik GmbH, Hauptverwaltung, Berlin
fxa — Eisenacher Karosseriefabrik Assman GmbH, Eisenach
fxo — C. G. Haenel, Waffen-und-Fahrrad-Fabrik, Suhl
fze — F. W. Holler, Waffenfabrik, Solingen
fzs — Heinrich Krieghoff, Waffenfabrik, Suhl

G.

ghf — Fritz Kiess & Co. Gmbh, Waffenfabrik, Suhl
gsb — Rheinmetall-Borsig AG, Louvain, Belgium
gsc — S.A. belge de Mécanique et de l'Armenente, Monceau-sur, Sambre, Belgium
guy — Werkzeugmaschinenfabrik Oerlikon, Buhrle & Co., Oerlikon, Zürich

H.

hew — F. Janecek, Prague
hhg — Rheinmetall-Borsig AG, Tegel, Berlin
hhv — Steyr-Daimler Puch AG, St. Valentin, Austria

J.

jhv — Metallwaren Waffen-u-Maschinenfabriken AG, Budapest
jkg — Kongl Ungar Staatl-Eisen-, Stahl- und Maschinenfabrik, Budapest
jij — Heeres Zeugamt, Ingoldstadt
jua — Danuvia Waffen- und Munitionsfabrik AG, Budapest
jwa — Manufacture d'Armes, Châtelleraut, France

K.

kfk — Dansk Industrie Syndicat, Copenhagen
kls — Steyr-Daimler Puch AG, Warsaw
ksb — Manufacture nationale d'Armes de Levallois, Paris
kur — Steyr-Daimler Puch AG, Warsaw
kwn — S.A. Fiat, Turin, Italy.

L.

lza — Mauser Werke AG, Werk Karlsruhe

M.

moc — Johann Springers Erben, Gewehrfabrikanten, Vienna
mpr — S.A. Hispano Suiza, Geneva
mrb — Aktiengesellschaft, formerly Skodawerke, Prague
myx — Rheinmetall-Borsig AG, Sömmerda, Thuringia

N.

nec — Waffenwerke Brünn AG, Prague
nhr — Rheinmetall-Borsig AG, Sömmerda
nyv — Rheinmetall Borsig AG, Unterlüss
nyw — Gustloff-Werke, Meiningen

42	Mauser Werke
S42	Mauser Werke
S42G	Mauser Werke
237	Mauser Werke
660	Mauser Werke
925	Mauser Werke

Note: Code numbers 42, S42, S42G, 237, 660 and 925 all indicate
Mauser Werke manufacture.

Arsenal or manufacturer's name, crest or trademark on a weapon
usually indicates manufacture before 1939

APPENDIX 5

German Small Arms Ammunition Manufacturers' Code

A.

ad — Patronen-Zündhütchen-u-Metallwarenfabrik, AG, Schoen-
     beck/Elbe

ak — Munitionsfabriken (formerly Sellier & Bellot), Prague

al — Deutsches Leucht-u-Signalmittelwerk, Dr. Feistel AG,
     Berlin

am — Gustloff Werke, Otto Eberhardt, Patronenfabrik, Hirten-
     berg, Niederdonau

an — Beuttenmuller & Cie., C., GmbH, Metallwarenfabrik,
     Bretten/Baden

ap — Gustloff Werke, Wuppertal-Romsdorf

asb — Deutsche Waffen-u-Munitionsfabriken AG, Berlin

asr — HAK (Hanseatisches Kettenwerk GmbH), Hamburg

auu — Patronenhülsen-u-Metallwarenfabrik AG Rokycany,
     Czechoslovakia

aux — Polte Werk, Magdeburg

auy — Polte Werk, Grüneberg

auz — Polte Werk, Arnstadt

axq — Erfurter Laden-Industrie, Erfurt

B.

bqt — Pyrotechnische Fabrik Eugen Müller, Vienna

byc — Klonne, Aug., Brückenbauanstalt, Dortmund

C.

cg — Finower Industrie GmbH, Finow, Mark Brandenburg

ch — Fabrique Nationale d'Armes de Guerre, Herstal, Liège,
     Belgium

czm — Gustav Genschow & Co. AG, Berlin

czo — Heeres Zeugamt, Geschosswerkstatt, Königsberg, Pr

D.

dbg — Dynamit AG (formerly Alfred Nobel & Co.), Duneberg
dma— Heeres-Munitionsanstalt, Geschosswerkstatt, Zeithain
dnf — Rheinisch-Westfalische Sprengstoff AG, Nuremberg
dnh — Rheinisch-Westfalische Sprengstoff AG, Durlach
dph — I.G. Farbenindustrie AG, Frankfurt/Main
dye — Erste Alpenländische Pyrotechnik, Ed. Pitschmann & Co.,
      Innsbruck

E.

ecc — Pyrotechnische Fabrik, Oskar Lunig, Mohringen
ecd — Earl Lippold, Pyrotechnische Fabrik, Wuppertal-Eberfelde
edg — J. A. Henckels Zwillingswerk, Solingen
edq — Deutsche Waffen-u-Munitionsfabriken AG, Lübeck-Schlut-
      up
eeg — H. Weirauch, Gewehr-u-Fahrradteilefabrik, Zella-Mehlis
      Thuringia
eel — Metallwarenfabrik (formerly H. Wissner AG), Hessen-
      Nassau
eem — Selve-Kronbiegel Dornheim AG, Sömmerda, Saxony
eeo — Deutsche Waffen-u-Munitionsfabriken AG, Posen
eom — H. Huck Metallwarenfabrik, Nuremberg
emp— Dynamit AG (formerly Alfred Nobel & Co.), Empelde

F.

fa — Mansfield AG, Hettstedt, Südharz
faa — Deutsche Waffen-u-Munitionsfabriken AG, Karlsruhe
fd — Stolberger Metallwerke AG, Stolberg
fde — Dynamit AG (formerly Alfred Nobel & Co.), Förde
fva — Draht-u-Metallwarenfabrik GmbH, Salzwedel

G.

gtb — J. F. Eisfeld, Pulver-und-Pyrotechnische Fabriken, Gun-
      tersberge

H.

has — Pulverfabrik Hasloch, Hasloch/Main
hgs — W. C. Gustav Burmester Pyrotechnische Fabrik-u-
      Signalmittelwerk, Hamburg
ham— Dynamit AG (formerly Alfred Nobel & Co.) Hamm
ha   — Metallwarenfabrik, Treunbritzen GmbH, Werke Seabaldus-
      hof
thg — Polte Werke, Duderstadt

J.

jtb — Tavavo, Geneva

K.

klb ·— Kiesselbach L. JFEs plant
krl — Dynamit AG Krümel
kry — Lignose Sprengstoffwerte GmbH, Kruppamuhle
kun — Werk-Kunigunde
kam— Hasag Eisen-u-Metallwerke GmbH, Werk Skarzysko-
      Damienna
kfg — Sarajevo State Arsenal
kyn — Astra, Fabrica Romana de Vagonne, Motoane Armament
      si Munitiuni, Brasow, Roumania
kye — Intreprinderile Metalurgie, Pumitta Voina Societate An-
      onima Romana, Fabrica de Armament, Brasow, Rouma-
      nia
kyp — Rumanisch-Deutsche Industrie-u-Handels AG, Bucharest
k   — Luch & Wagner, Suhl

L.

ldb — Deutsche Pyrotechnische Fabriken GmbH, Berlin
ldc — Deutsche Pyrotechnische Fabriken GmbH, Cleebronn
ldn — Deutsche Pyrotechnische Fabriken GmbH, Neumarkt/
      Oberpf
lge — Kugelfabrik Schulte & Co., Tente/Rhld
lkm — Munitionsfabriken (formerly Seller & Bellot), Prague

N.

nfx — RWS-Munitionsfabrik GmbH, Prague
nbe — Hasag, Eisen- und Metallwerke GmbH, Tschenstochau

P.

P-28   Polte, Magdeburg
P-120 Polte, Magdeburg
P-131 Polte, Magdeburg
P-315 Polte, Magdeburg
P-370 IX W Polte, Magdeburg
P-405 Polte, Magdeburg
P-490 Polte, Magdeburg
Pvf — Reichert, Vienna

Q.

qve — Carl Walther, Zella-Mehlis

V.

va   — Kabel-u-Metallwerke Neumeyer AG, Nuremburg

W.

wa   — Hugo Schneider AG, Lampenfabrik, Leipzig
wb   — Schneider AG, Berlin
wc   — Schneider AG, Meuschlwitz
wd   — Schneider AG, Taucha
we   — Schneider AG, Langeweisen
wf   — Hasag Eisen-u-Metallwerke GmbH, Kielce
wg   — Schneider AG, Attenburg
wh   — Schneider AG, Eisenach
wj   — Schneider AG, Oberweissbach
wk   — Schneider AG, Schlieben

Y.

Y   — Jagdpatronen-,   Zündhütchen-u-Metallwarenfabrik   AG,
        Budapest